PERSISTERS AND OPPORTUNISTS

IN AN ASSEMBLAGE OF

CORAL-REEF STARFISH

by

John C Paterson   B.Sc,

Zoology Department,

University of Queensland,

A thesis submitted according to the requirements
for the degree of Doctor of Philosophy
in the University of Queensland.

December, 1994

Contents Summary

CONTENTS 3

LIST OF TABLES 5

LIST OF FIGURES 6

ABSTRACT 7

ACKNOWLEDGMENTS 13

PREFACE 13

GENERAL INTRODUCTION 1

ASTEROID SPECIES PRESENT AT HERON REEF 12

HABITAT 32

POPULATION DENSITY 46

THE POPULATION SIZE STRUCTURE OF THE COMMONER SPECIES 65

SEXUAL REPRODUCTION 103

ASEXUAL REPRODUCTION 133

CONSTANCY OF MEAN SIZE 140

RELATIVE ABUNDANCE AND DIVERSITY 152

GENERAL DISCUSSION 162

REFERENCES CITED 183

CONTENTS

			Page
LIST OF TABLES			iv
LIST OF FIGURES			vi
ABSTRACT			vi
ACKNOWLEDGMENTS			xiii
PREFACE			xiii
CHAPTER	1	GENERAL INTRODUCTION	1
		Thesis to be defended	8
		Site of study	9
CHAPTER	2	SPECIES PRESENT	
	2.1	Introduction	12
	2.2	Methods	13
	2.3	Results	16
	2.4	Discussion	29
CHAPTER	3	HABITAT	
	3.1	Introduction	33
	3.2	Methods	36
	3.3	Results	37
	3.4	Discussion	41
CHAPTER	4	POPULATION DENSITY	
	4.1	Introduction	46
	4.2	Methods	47
	4.3	Results	49
	4.4	Discussion	61

CONTENTS (cont.)

			Page
CHAPTER	5	POPULATION SIZE STRUCTURE	
	5.1	Introduction	65
	5.2	Methods	66
	5.3	Results	67
	5.4	Discussion	99
CHAPTER	6	SEXUAL REPRODUCTION	
	6.1	Introduction	103
	6.2	Methods	104
	6.3	Results	106
	6.4	Discussion	127
CHAPTER	7	ASEXUAL REPRODUCTION	
	7.1	Introduction	133
	7.2	Methods	134
	7.3	Results	135
	7.4	Discussion	138
CHAPTER	8	CONSTANCY OF MEAN SIZE	
	8.1	Introduction	140
	8.2	Methods	142
	8.3	Results	142
	8.4	Discussion	149
CHAPTER	9	RELATIVE ABUNDANCE AND DIVERSITY	
	9.1	Introduction	152
	9.2	Methods	153
	9.3	Results	153
	9.4	Discussion	158
CHAPTER	10	GENERAL DISCUSSION	161
REFERENCES CITED			182

LIST OF TABLES

		Page
2.1	Asteroid species recorded from Heron Reef	17
3.1	Major zones in which each species occurs	38
3.2	Diet and location of each species	39
4.1	Density of each species	50
4.2	Density and patchiness of *Disasterina abnormalis*	52
5.1 - 5.10	Size data of species	68-95
5.11	Mean size of each species	98
6.1a - 6.8a	Spawning data of species	107-121
6.1b - 6.8b	Fisher's Exact test of responses	108-122
6.9	Type of reproduction of each species	126
8.1	Significance of variation in mean size	142
9.1	Relative abundance of each species	154

LIST OF FIGURES

		Page
1	Map of western end of Heron Island reef	11
4.1a	Number of individuals and sample area	51
4.1b	Number of species in each density category	51
4.2 - 4.15	Abundance distributions of species	53-60
5.1a - 5.10a	R-frequency distributions of species	69-96
5.1b - 5.10b	R/r-frequency distributions of species	69-96
5.1c - 5.10c	Relation between R and r of species	70-97
5.1d - 5.10d	Relation between R and R/r of species	70-97
6.1 - 6.8	Annual reproductive patterns of species	108-122
7.1	r and R/r variation in *Linckia multifora*	137
7.2	r and R/r variation in *Echinaster luzonicus*	137
8.1a	R variation in *Linckia multifora*	143
8.1b	R-frequency of *Linckia multifora* (May 1978)	143
8.1c	R-frequency of *Linckia multifora* (June 1979)	144
8.1d	R-frequency of *Linckia multifora* (July 1981)	144
8.2a	R variation in *Disasterina abnormalis*	145
8.2b	R-frequency of *Disasterina abnormalis* (May 1978)	145
8.2c	R-frequency of *Disasterina abnormalis* (Sep 1979)	146
8.2d	R-frequency of *Disasterina abnormalis* (Apr 1980)	146
8.3a	R variation in *Echinaster luzonicus*	147
8.3b	R-frequency of *Echinaster luzonicus* (May 1978)	147
8.3c	R-frequency of *Echinaster luzonicus* (Aug 1978)	148
8.3d	R-frequency of *Echinaster luzonicus* (Dec 1979)	148
9.1a	Number of individuals and rank abundance	155
9.1b	(log) % relative abundance and rank abundance	155
9.2a	Number of species and sample area	156
9.2b	Number of species and (log) sample area	156
9.3a	Shannon's Evenness and sample area	157
9.3b	Shannon's Evenness and (log) sample area	157

ABSTRACT

Population density, size-frequency and reproductive data on an assemblage of shallow water, coral-reef starfish (Asteroidea) were gathered over several years at Heron Reef. Heron Reef is a reef in the Capricorn Group at the southern end of the Great Barrier Reef. It has not been known to carry an outbreak of the corallivorous crown-of-thorns starfish (*Acanthaster planci*) and its coral cover is well developed. Specimens required primarily for size-frequency and reproductive analysis were collected by means of quadrats, general searches and intertidal traverses carried out at the western end of the reef. Most traverses included both reef flat and reef crest zones and all exposed starfish within a four meter width were collected for the length of the traverse. In addition, a selection of large and small, dead coral slabs were overturned and cryptic specimens located beneath these slabs were collected.

The finding of *Asteropsis carinifera*, *Dactylosaster cylindricus*, *Fromia elegans*, *Linckia multifora*, *Ophidiaster armatus*, *Ophidiaster lioderma*, *Ophidiaster robillardi*, *Asterina anomala*, *Disasterina abnormalis*, *Tegulaster emburyi*, *Mithrodia clavigera* and *Coscinasterias calamaria* represent new records for Heron Reef. This study has also provided the first record of the predominantly temperate species, *Coscinasterias calamaria* on a reef of the Great Barrier Reef. Essentially, the Heron Reef asteroid fauna is comprised of widely ranging West Pacific and Indo-West Pacific species plus a few species that appear endemic to the reefs of the Capricorn Group or are sub-tropical, rocky-shore species that have extended their ranges to include the southernmost reefs of the Great Barrier Reef.

It was found that Heron Reef carries a rich and diverse asteroid fauna and the linearity of the species : (log) area relationship indicates that additional species are still to be located. Of the 25 starfish species found on Heron Reef, 17

(*Asteropsis carinifera*, *Dactylosaster cylindricus*, *Fromia milleporella*, *Linckia laevigata*, *Nardoa novaecaledoniae*, *N. pauciforis*, *Ophidiaster confertus*, *O. granifer*, *O. lioderma*, *O. robillardi*, *Asterina anomala*, *A. burtoni*, *Disasterina abnormalis*, *D. leptalacantha*, *Tegulaster emburyi*, *Mithrodia clavigera* and *Coscinasterias calamaria*) were located only in intertidal regions. An additional three species (*Linckia guildingii*, *L. multifora* and *Echinaster luzonicus*) were found predominantly in intertidal regions but some specimens were located subtidally. *Culcita novaeguineae*, *Acanthaster planci*, *Fromia elegans*, *Gomophia egyptiaca* and *Neoferdina cumingi* were located predominantly in subtidal habitats, but have been recorded intertidally. While *Culcita novaeguineae*, *Fromia elegans*, *Gomophia egyptiaca*, *Linckia multifora* and *Echinaster luzonicus* were sometimes found at the base of the reef slope, they were never observed on the sea floor away from the reef. The preceding species can be regarded as coral-reef species and their distribution differs from that of species such as *Astropecten polyacanthus*, *Iconaster longimanus*, *Pentaceraster regulus*, *Leiaster leachi*, *Nardoa rosea*, *Ophidiaster armatus*, *Tamaria megaloplax*, *Echinaster stereosomus* and *Euretaster insignis* that are found in the deeper off-reef waters in the Heron Island region. The observation that most species of coral-reef starfish found on Heron Reef appear to be confined to the reef top (reef flat or reef crest) does not appear to have been noted previously.

The coral-reef asteroids found on Heron Reef showed some inter-specific variation with respect to diet but many species appeared to feed on epibenthic felt. The Heron Reef asteroids also showed some inter-specific variation with respect to habitat but some species occurred in exposed situations. Clear examples of niche specialisation (dietary or micro-habitat) are known only for the corallivores *Culcita novaeguineae* and *Acanthaster planci*. The Heron Reef asteroids did not appear to be resource limited.

Competitive interactions involving the Heron Reef asteroid assemblage were not observed during the five year period of the study as most species occurred with densities that were low. Indeed, the majority of species in this assemblage are considered to be rare or very rare. Even *Echinaster luzonicus*, the most abundant species, had an average density of only 16 specimens per hectare. Individuals of the more common species were patchily distributed (clumped). Only four species of starfish showed clear changes in density during the study period. These species were *Linckia multifora*, *Asterina burtoni*, *Disasterina abnormalis* and *Echinaster luzonicus*.

The small-bodied starfish *Disasterina abnormalis* occurred at an average density of over 8 individuals per square metre at one location on the northern reef crest but 100 metres away (still on the reef crest) its density was less than one individual per square metre. This region of high density of *Disasterina abnormalis* appeared to be confined to a narrow strip behind a rubble bank and this species was not found on 25 of the 72 traverses that were made. In this region, *Disasterina abnormalis* was highly clumped (at the metre square scale) in one sampling period and randomly distributed in another sampling period.

Juveniles of the relatively common, large-bodied, sexually reproducing asteroids *Linckia laevigata*, *Nardoa novaecaledoniae* and *N. pauciforis* were rare and their populations were adult-dominated throughout the study period. In all large-bodied species studied, distinct year classes were not observed in the population size structure and mortality was rarely observed. These species appear to possess low recruitment and low adult mortality. Juveniles were more common in the populations of *Linckia multifora*, *Asterina burtoni*, *Disasterina abnormalis* and *Echinaster luzonicus* and resulted from either sexual or asexual reproduction.

In the commoner of the large-bodied species, *Linckia guildingii*, *L. laevigata*, *Nardoa novaecaledoniae* and *Nardoa pauciforis*, little or negligible change in mean size was observed over the study period of five years indicating that these species are long-lived (persisters). In the same period,

Linckia multifora, *Disasterina abnormalis* and *Echinaster luzonicus* showed mean size variation that was highly significant, this variation being the result of periodicity in either sexual or asexual reproduction. Such species are short-lived (opportunists). In the other coral-reef asteroid species encountered, abundances were too low for statistically valid comparisons to be made.

Obvious changes in abundance due to either sexual or asexual recruitment, and significant changes in mean individual size were observed in the populations of *Linckia multifora*, *Disasterina abnormalis* and *Echinaster luzonicus*. While some recruitment and some change in abundance was noticed in both *Ophidiaster granifer* (parthenogenetic) and *Asterina burtoni* (hermaphroditic), no significant change occurred in the mean individual size of either species. On the other hand, *Linckia guildingii*, *Linckia laevigata*, *Nardoa novaecaledoniae* and *Nardoa pauciforis* exhibited only small changes in mean individual size and these species did not fluctuate greatly in abundance during the period of study. In these species, the population structure appeared to be adult-dominated and juveniles were encountered only rarely.

The relative stability of the size distributions of the common large-bodied species can be explained by assuming very slow growth of a predominant year class or a balance of recruitment and mortality within each of the species. It seems likely that a combination of both is involved. The paucity of juvenile asteroids, and the constancy of the size distributions in all the large bodied sexually reproducing species can be explained only by a life-history model which incorporates low adult mortality and includes the assumption of longevity.

Small sexual recruits of *Disasterina abnormalis* were relatively common in one highly localised area at Heron Reef, but high sexual recruitment was not observed in any of the other species. *Disasterina abnormalis* possessed small (non-yolky) sticky eggs that adhered to the substrate immediately following their release from the gonopores. All the remaining species possessed eggs that dispersed and underwent either planktotrophic or lecithotrophic larval development and no species were observed to brood larvae.

Culcita novaeguineae, *Acanthaster planci*, *Linckia guildingii* and *Linckia laevigata* were observed releasing eggs that contained little yolk and underwent planktotrophic development. *Fromia elegans*, *Gomophia egyptiaca*, *Nardoa novaecaledoniae*, *Nardoa pauciforis*, *Ophidiaster granifer* and *Echinaster luzonicus* were observed releasing eggs that contained large amounts of yolk and underwent lecithotrophic development. Specimens of both *Linckia multifora* and *Asterina burtoni* were injected regularly with 1-methyl adenine, but did not release gametes during the entire study. In addition to the species that demonstrated a sexual reproductive pattern, *Linckia guildingii*, *Linckia multifora*, *Ophidiaster robillardi* and *Echinaster luzonicus* reproduced asexually and exhibited comet stages while *Asterina anomala* and *Coscinasterias calamaria* reproduced asexually by binary fission.

With the exception of *Disasterina abnormalis*, all the species of starfish at Heron Reef either possessed a planktonic dispersive larval phase or did not reproduce sexually. The largest-bodied persistent species released planktotrophic eggs while the opportunist species were either lecithotrophic, hermaphroditic, parthenogenetic or solely asexually reproducing. No opportunistic species was observed to possess planktotrophic development.

Some species, namely *Disasterina abnormalis*, *Asterina burtoni*, *Ophidiaster granifer*, *Linckia multifora* and *Echinaster luzonicus*, could be regarded as opportunist species as they were characterised by possessing relatively abundant populations with relatively large fluctuations in mean individual size. These invariably small-bodied species demonstrated all of the typical opportunist characteristics including short life, high recruitment and high mortality.

Other species, namely *Culcita novaeguineae*, *Linckia laevigata*, *Linckia guildingii*, *Nardoa novaecaledoniae* and *Nardoa pauciforis* could be regarded as persistent species as proposed by Endean and Cameron (1990 a) and were characterised by less abundant populations with relatively smaller fluctuations in mean individual size than the opportunist species. These medium to large bodied species demonstrated all of the typical persister characteristics which include rarity, long life, low recruitment and low mortality. Despite searches over wide areas of different habitat, at different times over a 5 year period, a large proportion of the Heron Reef starfish species were sufficiently uncommon to preclude any analysis of either their relative abundance or size distributions. They have the attribute of rarity, which is characteristic of persisters, and are placed in this category pending further investigation. The persistent species in the Heron Reef asteroid assemblage appear to be recruitment limited.

The observed level of numerical and size-frequency stability in the persistent species is consistent with a model of community equilibrium. It is clear that mortality, dispersion, larval survival and settlement phenomena did not result in widely varying size structures or greatly differing adult numbers from one year to the next over a period of 5 years.

The results presented are in accord with the hypothesis of Endean and Cameron (1990 a) that complex, high diversity assemblages of coral-reef animals are characterised by a predominance of rare, long-lived species with relatively constant population sizes and size structures and a minority of opportunistic species characterised by fluctuating population sizes and size structures.

ACKNOWLEDGMENTS

I thank Dr Robert Endean for suggesting a basic plan and for supervising this study of coral-reef asteroids and for his helpful comments and continuing support during the extensive revision process. I thank the Heron Island Research Station for the use of its facilities.

I thank Jill Bricknell, Dr Ann Cameron, Ann Poulsen and Dr Russell Reichelt for reading and giving constructive comments on sections of this thesis. Their patience, insight and understanding have been, and always will be, appreciated greatly.

Over the period of this study, many people have assisted with data collection, logistic support and discussion. I thank them all, especially James Bricknell, Jill Bricknell, Steve Cook, Peter Gofton, Neil Gribble, Elizabeth McCaffrey, Norm Odgaard, Ann Poulsen, Russell Reichelt, Bill Stablum, Tim Stevens, Jude Westrup, Richard Willan and the students of ID211.

During the period of thesis revision, many others have assisted in the clarification of ideas by both their questions and discussion but I especially thank Lyndon Devantier, John Keesing, Brian Lassig, Hamish McCallum, Rob van Woesik and Leon Zann.

PREFACE

The work presented in this thesis is to the best of my knowledge and belief, original, except as acknowledged in the text. The material presented has not been submitted, either in whole or in part, for a degree at this or any other University.

....   John C Paterson
                                 22nd December 1994

CHAPTER 1

GENERAL INTRODUCTION

Coral reefs seem to defy many of the paradigms which characterise less complex biological communities. While there is general agreement that the biota of coral reefs exhibit high species diversity, some authors have characterised coral reef assemblages by selecting species with high population densities (Sale, 1974; 1976; 1977; 1984; Sale and Dybdahl, 1975; Connell, 1978). Other authors have included rarer species (Kohn, 1959; 1968; Den Boer, 1971; Grassle, 1973) and Endean and Cameron (1990 a) have emphasised the importance of the role of these rarer species and stated that rarity is virtually ignored in most ecological models of the coral reef ecosystem. They suggest that our understanding of coral-reef ecology is influenced strongly by the constraints of many of the analytical tools being used in reef studies. As a result they believe that most analyses have dealt primarily with species that are sufficiently numerous to provide statistically satisfactory numbers of records and that most studies have excluded rare species which, in fact comprise the majority of coral-reef species.

The complexity of coral reef ecosystems is not surprising given the great length of time that these ecosystems have been in existence. While the shallow water distribution of coral reefs has varied with the alternation of glacial and interglacial periods (Hays, Imbrie and Shackleton, 1976), in their broad biological form, coral reefs have existed since the Precambrian and reefs similar to present reefs have existed for around 50 million years (Newell, 1972). While stating that there is no general rule for coral-reef organisms, Endean and Cameron (1990 a) have suggested that the attribute of persistence possessed by most of the rarer species characterises the majority of coral-reef species and is responsible for both structuring and perpetuating this ecosystem. They regard the coral reef ecosystem as being an ordered and predictable system. However, other authors (Sale, 1977; 1991; Connell, 1978) have different views.

Sale (1991) regards reef fish communities as open non-equilibrial systems with living space determined in a random manner. Connell (1978) regards intermediate levels of disturbance as essential to the maintenance of diversity in this and other highly diverse and complex ecosystems. There has been much discussion of the meaning of stability (MacArthur, 1955; Dunbar, 1960; Leigh, 1965; May, 1972; Jacobs, 1974; Margalef, 1974; Goodman, 1975; Peters, 1976; Pimm, 1984).

Endean and Cameron (1990 a) have put forward the hypothesis that complex, high diversity assemblages of coral-reef animals are characterised by a preponderance of rare but long-lived species that they have termed persisters. These persistent species exhibit low recruitment, low adult mortality and relative constancy of adult population numbers and population structure. They occur in association with opportunist species that have high recruitment, a high adult mortality and varying adult population numbers and population structure. While individuals belonging to opportunist species are more abundantly represented than those belonging to persistent species, Endean and Cameron believe that the majority of species in the coral reef ecosystem are persistent species. This hypothesis has not been tested in the field.

As no general consensus relating to the organisation of coral reefs has been reached in the literature, the persister / opportunist distinction is examined in this thesis, rather than a deep analysis of the opposing views relating to stability. Events that are stochastic and unpredictable at one spatial or temporal scale may be predictable at another. In addition, the stability or otherwise of any system may be determined, amongst other things, by the particular set of species that is chosen to characterise the system.

The starfish fauna of coral reefs can be distinguished from the starfish fauna of surrounding waters (Endean, 1953; 1965) and coral-reef starfish may be regarded as an ecological entity. During studies of Queensland echinoderms, Endean (1953; 1957; 1961; 1965) found 18 species of starfish on Heron Reef. Although reference was made to the habitat, general abundance and biogeography of each of the species, no detailed study of the Heron Reef starfish assemblage was made. This study will compare a number of ecological parameters in several species of starfish occurring on this coral reef. The population stability of the less abundantly represented, persistent species will be contrasted with that of the more abundantly represented opportunistic species. For the purposes of this study, the population stability of each species refers to the constancy of its population size structure over time.

Clark and Rowe (1971) and Yamaguchi (1975 b) reviewed the geographic distribution of many coral-reef starfish. It is clear that specimens of some species are frequently encountered and appear to be relatively common while others are known from very few specimens and appear to be extremely rare. The ecological requirements of coral-reef starfish, as well as the role of both rare and common species, are not understood and it is not known whether rarity is a survival strategy, an abundance limit imposed by predators or a failure in competitive ability of a species on its path to extinction. These problems have not been addressed for asteroids or any other taxonomic group within the highly diverse and complex coral reef ecosystem.

It has been suggested that longevity may characterise species of predictable environments (Frank, 1968; Grassle, 1973) or species with unpredictable pre-reproductive survival (Ebert, 1982; Goodman, 1974; Murphy, 1968). Several authors (Frank, 1969; Grassle, 1973; Ebert, 1982) have found many coral-reef animals to be long lived and Endean and Cameron (1990 a) regard the long-term persistence of individuals at given sites as an ordering phenomenon in the coral reef ecosystem. Little information is available on the longevity of coral-reef starfish. Ebert (1983), Kenchington (1976), Cameron and Endean (1982) and Endean and Cameron (1990 b), believe that *Acanthaster planci* is a long lived species, but Lucas (1984) suggested individual senescence in this species at an age of approximately five years. Stump and Lucas (1990) reported a linear growth pattern in aboral spine ossicles of this species which supported this suggestion, however the maximum age of this species has now been re-evaluated to at least 12-15 years (R.Stump, Ph.D. thesis). Yamaguchi and Lucas (1984) demonstrated a short lived population structure in the small and cryptic starfish *Ophidiaster granifer*, but little is known of the longevity of other species of coral-reef starfish.

The severe effects of *Acanthaster planci* predation are well documented (Chesher, 1969 a,b; Endean, 1969) and the change in coral population structure following an *A. planci* population outbreak was reported by Cameron, Endean and Devantier (1991). Moran (1986) has compiled a bibliography on the *Acanthaster planci* population outbreak phenomenon. Research on temperate starfish species that undergo population outbreaks has been reviewed by Loosanoff (1961).

Little is known of the other coral-reef starfish species, and the reproductive patterns, population stability and diversity of starfish assemblages on reefs that have not carried population outbreaks of *Acanthaster planci* are poorly understood. Heron Reef is such a reef. It is a Marine National Park and is situated near the southern end of the Great Barrier Reef.

It should be appreciated that the number of species recorded in any study is determined by both the spatial and temporal scales of sampling as well as by the distribution and composition of the species in the assemblage (species richness or diversity). To allow some degree of standardisation for collection effort, the rate at which the number of species in a sample increases with area of the sample (the species-area relationship) and the range of abundances within this assemblage (species relative abundance) have been chosen as a more representative measure of species richness than the total number of species. The species-area relationship, relative species abundance, and constancy of numbers and mean size (population stability) of coral-reef starfish are unknown elsewhere.

The fact that this study was undertaken on a reef that appeared to have low starfish abundance (and was not known to have carried an *Acanthaster* outbreak) generally precluded small-scale analyses that were dependent on high population densities. Large-scale traverse sampling is analogous to manta tows that have been used to monitor populations of *Acanthaster*. This scale of sampling is useful to establish a general pattern of starfish abundance, but is not capable of providing detailed data on either microhabitat partitioning or small-scale abundance. It should provide a basis for future comparison with data from Heron and other reefs.

Sexual reproductive patterns have been studied in some of the coral-reef starfish species known to occur throughout the Indo-West Pacific. Most of these studies have been conducted on reefs that are known to have carried population outbreaks of *Acanthaster planci* (Yamaguchi, 1973 a,b; 1974; 1975 b; 1977 a; Yamaguchi and Lucas, 1984). In addition to providing reproductive data for these species from a reef that does not undergo such population outbreaks, this study will examine the reproductive patterns of previously unstudied species. When the timing and extent of sexual reproduction along with the type of larval development exhibited by the various species studied are correlated, inferences can be drawn regarding the reproductive effort and dispersal capacity of each species involved. Endean and Cameron (1990 a) have suggested that opportunists and persisters are basically different with respect to their rates of recruitment, and a pattern should emerge when data on reproduction and population structure for a number of coral-reef starfish species are compared.

Several species of coral-reef starfish are known to exhibit asexual reproduction. The extent of asexual reproduction in the population maintenance of each species is an indication of the adaptive significance of this low-dispersal reproductive strategy. Many authors have commented on the role that may be played by this form of reproduction (Rideout, 1978; Yamaguchi, 1975 b; Ottesen and Lucas, 1982; Yamaguchi and Lucas, 1984) and Endean and Cameron (1990 a) have suggested that this mode of reproduction may assist species to withstand disturbance. Most species of starfish cannot reproduce asexually but are still capable of great powers of regeneration. Missing limbs in species that do not reproduce asexually may indicate sub-lethal predation.

Recruitment, migration and mortality ultimately determine the spatial and temporal distributions of the starfish populations in the Heron Reef assemblage. There is a distinction to be drawn between reproduction and recruitment as well as between predation and mortality. Recruitment is a process that is complete only when an offspring reaches maturity and reproduces itself. Similarly, predation may only be sub-lethal and autotomised limbs may be regenerated or may become asexual recruits. Starfish mortality occurs only when all fragments of a starfish have died. For logistical reasons, it was decided not to examine potential predators in this study. Likewise, a detailed examination of larval settlement processes was not undertaken. Migration of starfish is poorly understood as there is considerable difficulty in relocating tagged specimens particularly in autotomous species.

However, the interaction between the major determinants of population size mentioned above will influence the size-frequency distributions of each species. These distributions will be compared over time at Heron Reef and with size-frequency data from other localities. Mean individual size will vary with periods of recruitment and mortality, and size-frequency distributions that are constant over a study period of several years will suggest stability within the age structure. Alternately, such a finding could reflect the apparently static nature of a long-lived species when observed on a comparatively short time scale, even one of several years. However, study of the latter alternative could not be pursued beyond the time frame of this study which embraced five years.

Knowledge of the spatial pattern, fecundity and population dynamics of each of the coral-reef starfish species represented is essential to an understanding of the stability or otherwise of the populations of species comprising the coral-reef asteroid assemblages of Indo-West Pacific reefs. This knowledge is also essential to an understanding of outbreak phenomena, such as population outbreaks of *Acanthaster planci*. The obtaining of comparative distribution and reproductive data on many starfish species from Heron Reef will clarify the factors that influence diversity and stability within this assemblage.

With these broad aims in mind, this study focused on Heron Reef and sought answers to the following questions:

What starfish species are present at Heron Reef?
What is the spatial pattern for each species?
What is the population structure of each species?
What is the reproductive mode of each species?
Is the mean individual size stable for each species?
How is abundance distributed within this assemblage?

Thesis to be defended

In this study of the shallow-water asteroid assemblage of Heron Island reef, an Indo-West Pacific coral reef that has not been known to carry an outbreak of *Acanthaster planci* and hence can be regarded as a reef that has not been subject to a major disturbance at least in the immediate past, the thesis to be defended is:

1. The asteroid assemblage is comprised of numerous persistent species and a smaller number of opportunistic species.

2. The persistent species are relatively uncommon (rare) and possess relatively stable population densities and population size structures and have low rates of recruitment.

3. The opportunistic species exhibit localised high density, significant population fluctuations and are characterised by high recruitment (either sexual or asexual).

Site of study

Heron Reef (23° 27' S, 151° 57' E) lies in the Capricorn Group which is towards the southern end of the Great Barrier Reef. It is a lagoonal platform reef with a vegetated cay at its western end (Figure 1). The cay supports a tourist resort and research station. Heron Reef has been zoned as Marine Park A within the Capricornia Section of the Great Barrier Reef Marine Park, and prior to this was protected, from over-collection, by a regulation of Queensland State Fisheries. The western end of the reef is easily accessible from the cay but access to the eastern end requires the use of a small boat.

The major habitat zones used in the present study are described in detail by Jell and Flood (1978).

These zones are:
1. Reef flat (with lagoon)
2. Reef crest or reef rim
3. Reef slope
4. Off-reef floor

At the western end of Heron Reef, where studies were made, the reef flat is the sub-tidal habitat nearest to the cay. It is chiefly comprised of dead and living coral clumps which vary in size from a few centimetres in diameter to dead coral boulders or living micro-atolls with diameters of several metres. The dead coral clumps can, at certain times of the year, be obscured by a prolific growth of algae. The chief physical parameter that separates the reef flat from the lagoon is the water depth at low water spring tides. The water depth can vary from less than half a metre at the western end of the reef where sedimentation is great to more than a metre at its transition into lagoon east of the cay. The lagoon is up to six meters in depth at Heron Reef and has scattered coral outcrops which may reach the surface. It is regarded as an extension of the reef flat for the purposes of this study. At the innermost part of the reef flat (adjacent to the cay) a series of strata composed of cemented sand and coral fragments occurs. The strata are called beachrock.

The reef crest is the outer region of intertidal coral growth and is shallower than the previous zone. It is the most turbulent of all coral-reef zones being exposed to direct wave action at all stages of the tide. It has little fine sediment other than that which is trapped within the algal turf and which has accumulated under boulders. Living coral growth is usually low in profile and the general substrate is comprised of cemented reef rock strewn with broken coralline material. This material ranges in size from single coral fragments which are a few centimetres in diameter, to large boulders that are greater than two meters in diameter.

The reef slope is subtidal and supports extensive coral growth to a depth of approximately 20 meters. The coral growth tapers off to almost negligible coral cover at a depth of approximately 30 meters where the slope merges with the off-reef floor. This transition may be sudden on some reefs which possess almost vertical reef slopes, but at Heron Reef the transition is gradual. This zone is less physically controlled than are the previous zones. After periods of severe swell there may be areas of broken coral colonies but generally, as depth increases, the direct effect of wave action decreases. The substrate is of poorly sorted sediments as well as living and dead coral colonies, together with their epibiota.

The off-reef floor between Heron Reef and the adjacent reefs is over 40 meters deep and in places supports a well developed fauna of alcyonarians and solitary hard corals along with their associated epibiota. The off-reef floor is the deepest of the reef zones and provides habitats that are clearly different from the shallow water habitats provided by the other three zones. The sediment found on the off-reef floor is varied and its composition is dependent on currents as well as on surge effects during heavy wave action.

Figure 1.

Map of western end of Heron Reef. (After Jell and Flood, 1978)

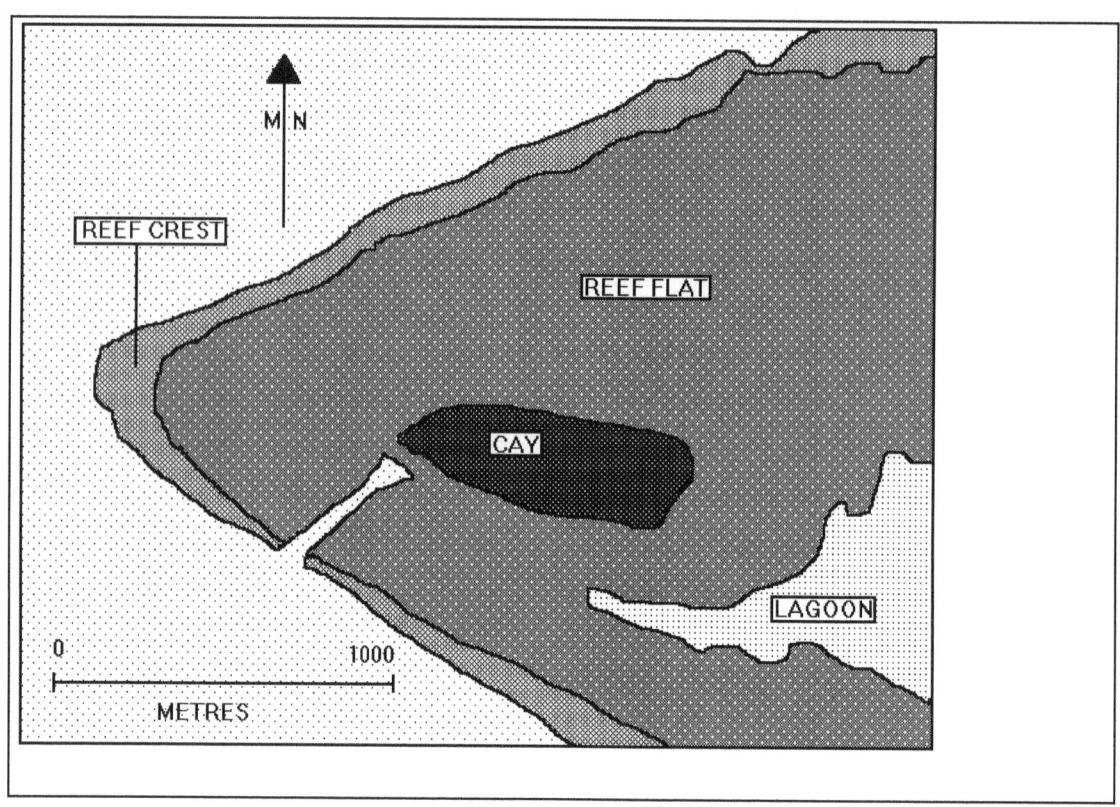

CHAPTER 2

ASTEROID SPECIES PRESENT AT HERON REEF

2.1 Introduction

An initial study of the echinoderms of the Great Barrier Reef was conducted by H.L.Clark during a visit to the Murray Islands in 1913 (Clark, 1921). This work was followed by that of A.A.Livingstone during the Great Barrier Reef Expedition (Livingstone, 1932) and that of Gibbs, Clark and Clark (1976). Two monographs dealing with the Australian echinoderm fauna were compiled by H.L.Clark (Clark, 1938; 1946). Extensive biogeographical studies of Queensland echinoderms were undertaken by Endean (1953; 1956; 1957; 1961; 1965) and many of the records therein relate to Heron Island asteroids.

In the Indian Ocean, a detailed account of the echinoderm species present in West Australian waters was provided by Marsh (1976). Elsewhere in the Indian Ocean, the asteroid (starfish) fauna has been studied at Mozambique (Jangoux, 1972 a; Walenkamp, 1990), South Africa (Thandar, 1989), Somalia (Tortonese, 1980), the Gulf of Suez (James and Pearce, 1969), the Red Sea (Clark, 1967 a; Tortonese, 1960, 1977, 1979), the Arabian Gulf (Price, 1981), the Iranian Gulf (Mortensen, 1940), India (Koehler, 1910; James, 1973), the Andaman and Nicobar Islands (Julka and Das, 1978) and the Maldive Islands (Clark and Spencer-Davis, 1966; Jangoux and Aziz, 1985).

In the Pacific Ocean, the starfish fauna has been studied in China (Liao, 1980), Hong Kong (Clark, 1982), Taiwan (Chao and Chang, 1989), the Philippines (Fisher, 1919; Domantay, 1972; De Celis, 1980), the Ryukyu Islands (Hayashi, 1938 a), the Ogasawara Islands (Hayashi, 1938 b), the Caroline Islands (Hayashi, 1938 c; Grosenbaugh, 1981; Marsh, 1977; Oguro, 1984), the Mariana Islands (Yamaguchi, 1975 b; Kerr *et al.*, 1992), the Marshall Islands (Clark, 1952), Indonesia (Guille and Jangoux, 1978; Jangoux, 1978), New Caledonia (Jangoux, 1984), Tonga (Clark, 1931), South East Polynesia (Marsh, 1974), Hawaii (Fisher, 1906; Ely, 1942) and the general North Pacific region (Fisher, 1911, 1925). The geographical distribution of the shallow water species was reviewed by Clark and Rowe (1971).

There have been many taxonomic revisions within the Asteroidea. The works of Baker and Marsh (1974), Blake (1979; 1980; 1981; 1983; 1990), Jangoux (1972 b; 1980), Pope and Rowe (1977), Rowe (1977) and Marsh (1991) have included species of coral-reef starfish. All previous revisions were summarised in the specific descriptions and keys to the asteroid species provided by Clark and Rowe (1971).

2.2 Methods

Specimens of several species of starfish were required primarily for size-frequency and reproductive analysis. Sampling methods were chosen so as to ensure that the sample sizes were sufficient to allow statistical analysis of size-frequency and reproductive data in a reasonable number of species. Starfish were collected by means of quadrats, general searches and on traverses that were conducted primarily at the western end of Heron Reef (Figure 1). On Heron Reef, traverses ran between the cay and the reef crest (0.5 to 2 kilometres apart) and also between two points both on the reef crest (0.5 to 6 kilometres apart). Because the primary purpose of sampling was the collection of size-frequency and reproductive data, the traverses were not stratified with respect to habitat. Traverses were neither systematic nor random and most traverses included both reef flat and reef crest zones. All exposed starfish within a four meter width were collected for the length of the traverse. In addition to the collection of exposed starfish, a selection of large and small, dead coral slabs were overturned and cryptic specimens located beneath these slabs were collected. The lagoon and its adjacent coral pools were not sampled by traverse because of the difficulty in traversing this habitat.

All traverses were conducted within two hours of low tide, during the period of spring tides (full or new moon). When the water over the reef flat and reef crest was any deeper than this or under adverse weather conditions it was difficult to locate smaller starfish. Specimens that were required for reproductive studies were collected during general searches at these times but these specimens were not included in either the abundance or size-frequency data because this would have been biased towards the more visible species and individuals.

In total, 72 overlapping, intertidal traverses were conducted during the period from May 1978 to December 1982. The total area sampled by these traverses was approximately 120 hectares (1.2 square kilometres) which is equivalent to about five percent of the shallow-water, reef area of Heron Reef. The mean traverse length was just over four kilometres.

Cryptic species were also sampled using metre square quadrats in particular areas where previous traverse sampling had shown that starfish abundance was relatively high. These samples provided data for starfish present on a very small area of the reef crest. These quadrat samples cannot be regarded as random and they are not typical of the reef crest in general. The reef crest zone is extremely variable and spatial heterogeneity (patchiness) appeared to be highly dependent on the scale of sampling. These quadrat samples were undertaken to obtain estimates of the starfish density in these localised patches.

Subtidal specimens of starfish were collected on the reef slope and off-reef floor by the use of SCUBA. These subtidal samples were not used to determine subtidal starfish density because limitations in underwater visibility would have resulted in the underestimation of all starfish abundances. Detailed quadrat sampling would not have been directly comparable with intertidal traverse data and such sampling was not considered appropriate given the logistical constraints of extensive sampling using SCUBA. The off-reef floor was only rarely sampled and the species that occur in this habitat may be much more abundant than is apparent from the results obtained.

All starfish were identified, measured and placed along with conspecifics in glass aquaria at the Heron Island Research Station. Specimens were identified by reference to Clark and Rowe (1971). Specimens were also compared with their original descriptions where necessary. An examination of the specimens with a stereoscopic microscope was sufficient to distinguish all species. Juvenile identification was possible in all cases by reference to Clark (1921), Yamaguchi and Lucas (1984) or Yamaguchi (1973 a, 1973 b, 1974, 1977 a).

All individuals not required for taxonomic study were released in habitats similar to those where they were found. Specimens of all species studied were photographed live and some were preserved in alcohol. These are housed in the Department of Zoology, University of Queensland.

Throughout this thesis, unless some ecological parameter is given higher priority temporarily, the sequence in which species appear in tables is determined by their systematic position. The families are sequenced according to Blake (1979, 1980, 1981, 1987, 1990) and the classification of Clark and Rowe (1971). The genera are sequenced alphabetically within families.

2.3 Results

The species listed in Table 2.1 have either been recorded previously from Heron Reef, or were found in the present study and represent new records for the locality (marked with "*"). The species included are all those that occur either on the reef top (reef flat and reef crest) or on the reef slope extending to a depth of approximately 30 metres. At Heron Reef this is approximately the depth where the substrate of predominantly live coral or coral rubble changes to the finer sediments of the off-reef floor. Coral-reef species that do not appear to occur on the off-reef floor are marked "+".

Table 2.1 Asteroid species recorded from Heron Reef.

Astropectinidae
Astropecten polyacanthus Muller and Troschel,1842

Goniasteridae
Iconaster longimanus (Mobius, 1859)   *

Oreasteridae
Culcita novaeguineae Muller and Troschel,1842   +

Acanthasteridae
Acanthaster planci (Linnaeus,1758)   +

Asteropseidae
Asteropsis carinifera (Lamarck,1816)   *+

Ophidiasteridae
Dactylosaster cylindricus (Lamarck,1816)   *+
Fromia elegans Clark,1921   *+
Fromia milleporella (Lamarck,1816)   +
Gomophia egyptiaca Gray,1840   +
Linckia guildingii Gray,1840   +
Linckia laevigata (Linnaeus,1758)   +
Linckia multifora (Lamarck,1816)   *+
Nardoa novaecaledoniae (Perrier,1875)   +
Nardoa pauciforis (von Martens,1866)   +
Nardoa rosea Clark,1921
Neoferdina cumingi (Gray,1840)   +
Ophidiaster armatus Koehler,1910   *
Ophidiaster confertus Clark,1916
Ophidiaster granifer Lutken,1871   +
Ophidiaster lioderma Clark,1921   *+
Ophidiaster robillardi de Loriel,1885   *+
Ophidiaster watsoni (Livingstone,1936)   +
Tamaria megaloplax (Bell,1884)   *

Asterinidae
Anseropoda rosacea (Lamarck,1816)
Asterina anomala Clark,1921   *+
Asterina burtoni Gray,1840   +
Disasterina abnormalis Perrier,1876   *+
Disasterina leptalacantha (Clark,1916)   +
Tegulaster emburyi Livingstone,1933   *+

Mithrodiidae
Mithrodia clavigera (Lamarck,1816)   *+

```
Echinasteridae
Echinaster luzonicus (Gray,1840)                          +
Echinaster stereosomus Fisher,1913                        *

Asteriidae
Coscinasterias calamaria (Gray,1840)                      *

*  new record for Heron Reef      +  coral-reef species
```

In addition to the preceding species, *Anthenea aspera*, *Stellaster equestris*, *Metrodira subulata* and *Acanthaster brevispinus* were recorded from the area by Bennett (1958). These species were dredged from a depth of 45 meters east of Wistari Reef and were not directly associated with any coral-reef habitat. *Halityle regularis* was recorded by Baker and Marsh (1974) and *Andora popei* was recorded by Rowe (1977) from the off-reef floor near Heron Reef. *Pentaceraster regulus* and *Euretaster insignis* were also observed on the off-reef floor.

The following brief notes relate to the species of starfish that have been located at Heron Reef (on the reef flat, reef crest or reef slope) either in this study or by previous workers.

Family Astropectinidae

Astropecten polyacanthus Muller and Troschel,1842

This species of starfish is not restricted to coral reefs, but occurs also in sandy areas along the east coast of the Australian mainland. It was not common but specimens were found during this study in the deeper waters of the off-reef floor. It is recognised by the many conspicuous sharp spines along the body margin. The tube feet do not possess suckers at their tips. It has been found on a sandy spit at Heron Island Reef by Endean (1965).

Family Goniasteridae

Iconaster longimanus (Mobius,1859)

This orange and white patterned starfish is immediately recognised by its long tapering arms. It was not common at either Heron or the adjacent Wistari Reef, but specimens were located during this study in about 20 metres of water on the deeper parts of the reef slope. They were usually associated with coral rubble. Some specimens that were collected had recently lost one arm.

Family Oreasteridae

Culcita novaeguineae Muller and Troschel,1842

The juveniles of this species (R less than 70 mm) look quite different from adults. This starfish is most commonly encountered on the reef flat although it occurs also on the reef crest. Its greatest abundance may be at the base of the reef slope or in the deeper coral pools adjacent to the lagoon. This large and conspicuous species was not common on the traverses at Heron Reef during the period of this study.

Family Acanthasteridae

Acanthaster planci (Linnaeus,1758)

This well known species was uncommon at Heron and the adjacent Wistari Reef during the period of the present study. Only five subtidal adults and one juvenile specimen were encountered. Endean (1961) recorded a single specimen from a pool near the reef crest at Heron Reef.

Family Asteropseidae

Asteropsis carinifera (Lamarck,1816)

This species is not common at the southern end of the Great Barrier Reef. It has been recorded as common at Mer in the

Murray Islands (Clark, 1921). During this study, three specimens were encountered on the reef crest at Heron Reef.

Family Ophidiasteridae

Dactylosaster cylindricus (Lamarck,1816)

During this study, a single specimen was located on the reef crest at Heron Reef. Few specimens of this species have been found on the Great Barrier Reef or elsewhere throughout its range. This species can be distinguished from others in this family by the presence of only a few small granules in the centre of each plate of the body. The remaining granules are concealed by a skin-like membrane. It occurs on the rocky reefs off southern Queensland more frequently than it does at Heron Reef.

Fromia milleporella (Lamarck,1816)

One specimen was found on the reef crest. Endean (1956) found two specimens under boulders on the reef crest.

Fromia elegans Clark,1921

At Heron Reef, this starfish is relatively common in the reef slope zone. Most specimens have five even arms, but specimens with four and six arms were not uncommon. This species was found also on the reef crest lying exposed in small pools, and on the sand at the base of the reef slope in 20 metres of water.

Gomophia egyptiaca Gray,1840

At Heron Reef, the only individuals encountered, during this study, were coloured purple and brown with pink tips to the tubercles which cover the aboral surface of the body. Specimens of this species were usually found either concealed under boulders on the reef crest or crawling amongst dead coral rubble on the reef slope. This species is not common at Heron Reef. Of the small number of intertidal specimens collected, two were found in close proximity. Endean (1965) found only two specimens on the reef flat at Heron Reef.

Linckia guildingii Gray,1840

The Grey *Linckia*, while not as common as *L. laevigata* or *L. multifora*, is encountered frequently on reefs of the Great Barrier Reef. The grey coloration conceals the animal when crawling over dead coral clumps which are covered by filamentous algae, but the animal is conspicuous when on coral sand. Although the adult starfish is uniform grey in colour, juveniles are mottled white, grey and purple, and do not lose this appearance until a size of about 80 mm arm radius is attained.

Linckia laevigata (Linnaeus,1758)

The Blue *Linckia* inhabits intertidal reef areas throughout the Indo-West Pacific region. It attains a large size (arm radius 180 mm), is brightly coloured, and is usually found lying unconcealed on or near coral clumps in the reef flat. It can be found also on the reef crest, lying either exposed on the algal rim or partially hidden under coral boulders in the rubble zone. There is very little colour variation within this species on the Great Barrier Reef. The most frequent number of arms is five although arm number ranges from three to seven. The extremes are rare.

Linckia multifora (Lamarck,1816)

This species is usually found with one or more arms missing, these having been autotomised. The maximum size that this animal attains at Heron Reef is about 100 mm arm radius, but most specimens are approximately one-third this size. Sometimes the starfish will be found crawling in the open across the reef crest but more often it will be found under boulders. The specimens which occur under boulders are usually smaller and lighter in colour and do not have the brown coloration which is found in those that have adopted an exposed existence. The most common number of arms is six but the number varies between three and eight. It is unusual to find a specimen with all arms of equal length.

Occasionally specimens are found that do not belong clearly to either *Linckia laevigata* or *Linckia multifora*. These specimens are blue in colour but have pointed arms and show evidence of recent autotomous reproduction. There is a small row of granules between the furrow spines. One blue comet form has been found during this study. Because of their general morphology, these specimens have not been regarded as *Linckia laevigata*, but as colour variations of *Linckia multifora*.

Nardoa novaecaledoniae (Perrier, 1875)

The two common species of *Nardoa* appear quite similar in overall appearance and differ in the arrangement of the plates which cover the arms. In *Nardoa novaecaledoniae* these plates are abruptly reduced in size in the outer one-third of each arm.

Nardoa pauciforis (von Martens, 1866)

This starfish is slightly less common than the previous species but is not hard to find on Heron Reef. It occurs more commonly on the reef flat than on the reef crest but it can be overlooked in this habitat as both *N. pauciforis* and *N. novaecaledoniae* blend well with the background of living and dead coral. The animals are most conspicuous when crawling over the sand between coral clumps. The average individual size of this species is slightly larger than that of *N. novaecaledoniae*. Also, the arms are usually longer relative to the body than in *N. novaecaledoniae*. A diagnostic feature of *N. pauciforis* is the absence of an abrupt change in the size of the plates towards the outer one-third of the arms.

Nardoa rosea Clark,1921

This species is more frequently encountered in the deeper parts of the reef slope (20 meters) than on the top of the reef at Heron Reef, but is not common in any of these habitats. It is a beautiful starfish with an average size of 90 mm arm radius.

Neoferdina cumingi (Gray,1840)

This starfish is not encountered often on the reef top at Heron Reef, but is occasionally seen when diving on the reef slope. There is great variation in the number and pattern of the red spots which are conspicuous along the arms.

Ophidiaster armatus Koehler,1910

All members of the genus *Ophidiaster* possess four rows of papular areas on both sides of every arm, a total of eight rows per arm. Papulae are the respiratory organs and occur in groups of between five and twenty, each appearing as a small, transparent projection through the outside body wall. The extent to which each papula is extended is dependent greatly on the water conditions.

O. armatus is readily recognisable by its dark coloration, tapering arms and by the coarse feel of the animal due to the very rough granulation of its skin. This species is found in low numbers, mainly at the base of the reef slope at Heron Reef.

Ophidiaster granifer Lutken,1871

This species possesses the tapering arms and uneven granulation of the previous species, but it is easily distinguished by its general coloration, smaller size (25 mm) and shorter arms relative to the diameter of the disc. Specimens of this species are usually encountered under boulders on the reef crest where they occur with moderate abundance. They are always cryptic in their habits.

Ophidiaster lioderma Clark, 1921

This moderate sized starfish (R=100 mm) is very rare indeed having been found on two known occasions only, in two localities which are far apart on the Great Barrier Reef. The original specimen was discovered by H.L.Clark when he visited the northern end of the reef and was based at Murray Island in Torres Strait at the turn of the century. During this study, a further specimen was located on the reef crest at Heron Reef and is now housed in the West Australian Museum.

This species is a medium-brown in colour and can be readily identified by the skin covered body which possesses microscopic granulation. The only other member of this family which has a covering of thick skin is *Leiaster leachi* but this species has no surface granulation whatsoever and is brightly coloured.

Ophidiaster confertus Clark, 1916

Four specimens of this species were located on the reef crest at Heron Reef. This species which grows up to 160 mm arm radius occurs more commonly on the New South Wales coast than at the southern end of the Great Barrier Reef (Clark, 1946).

Ophidiaster robillardi de Loriel, 1885

This species occurs in moderate abundance in patches at Heron Island. The extreme patchiness of the distribution and abundance of this species is attributable to low dispersion associated with asexual reproduction. The average size of specimens is 35 mm arm radius and about ten percent of the specimens encountered were comet forms resulting from autotomous reproduction.

Ophidiaster watsoni Livingstone,1936

Gomophia egyptiaca and *Ophidiaster watsoni* are very similar and may be conspecific. Endean (1956) found one specimen of *O. watsoni* under a boulder on the reef edge at Heron Island.

Tamaria megaloplax (Bell,1884)

This species is found in the deeper waters, on sand near the base of the reef slope at Heron Reef. It occurs much more commonly on rocky reefs in south-east Queensland, than it does on the Great Barrier Reef. The average size of specimens found in south-east Queensland is about 100 mm arm radius. The specimens show considerable variation in the degree of roundness of the plates on the arms. This genus is characterised by having only three parallel rows of papular groups on both side of each arm, unlike *Ophidiaster* which has four, and *Hacelia* which has five rows.

Family Asterinidae

Anseropoda rosacea (Lamarck, 1816)

A single specimen of this species was found on sand in a reef-crest pool at Heron Reef by Endean (1956).

Asterina anomala Clark,1921

This small starfish is usually hard to find as the maximum size of individuals found on the Great Barrier Reef is about 5 mm arm radius. The bright coloration is of little help in finding this species as the boulders under which it occurs are encrusted usually with other brightly coloured invertebrates such as sponges and ascidians. This species is probably much more common than it appears to be but its small size makes sampling extremely difficult.

The usual number of arms in this species is seven. Half of these are normally regenerating as this species reproduces by binary fission. In this process, the animal divides into two and both sides regenerate the missing arms. If the regeneration has not proceeded very far then three or four adjacent ambulacral grooves will not extend much beyond the mouth.

Asterina burtoni Gray,1840

The taxonomic positions of this and of the preceding species are not clear. While the coloration of *Asterina burtoni* is quite variable, ranging from grey, through green to red or purple, it does not exhibit the multi-coloured pattern possessed by the previous species. *A. burtoni* does not reproduce by fission at Heron Reef and consequently most specimens have five arms of equal length. The average size of specimens is 13 mm arm radius.

Disasterina abnormalis Perrier,1876

This species has been recorded at a few localities along the Great Barrier Reef, and also in Indonesia as well as in the South Pacific. When alive, the animal is covered by a relatively thick skin which conceals the underlying plates. Many of these plates bear some very short rounded spines but it is not possible to discern the diagnostic characters of this species unless the specimen is preserved and then dried.

At Heron Reef at the southern end of The Great Barrier Reef, this is the most abundant starfish found on the top of the reef. It lives amongst the broken coral rubble on the innermost portion of the reef crest. The average size of specimens is 15 mm arm radius, but this size varies with periods of growth and with recruitment of juveniles to the population.

Disasterina leptalacantha (Clark, 1916)

This close relative of the preceding species grows to the same size, but is known only from the Capricorn Group at the southern end of the Great Barrier Reef. The difference between these two species is unmistakable as *Disasterina leptalacantha* possesses very long, extremely thin spines along the body margin, but in life these may be folded upwards against the side of the body and are overlooked easily. The coloration of this species is different from that of the previous one and the arms are also slightly longer. The reason for the apparent limited distribution of this species is unknown.

The main habitat of this species is amongst the broken slabs of beachrock at low tide level. It is not common but specimens will be found either adhering to the underside of the rocks or amongst the sand immediately under the rocks.

Tegulaster emburyi Livingstone, 1933

During this study, one specimen of this species was located on the reef crest at Heron Reef. The only other known specimen of this species was found at North-West Island, also in the Capricorn Group. Both specimens were found under a dead coral boulder in the reef crest zone. This species is exceedingly rare and may also be highly restricted in its geographic range. Both known specimens were just under 20 mm in arm radius.

Family Mithrodiidae
Mithrodia clavigera (Lamarck,1816)

During this study, one specimen was located at Heron Reef, but it did not occur within the intertidal traverses. It was located on the reef crest in December 1984. The species has been found elsewhere in the South Pacific but is uncommon.

Family Echinasteridae
Echinaster luzonicus (Gray,1840)

This starfish ranges from almost black, through red, to speckled orange and black in coloration. Specimens with all arms of equal length are not common as this species reproduces by means of autotomy, and comet forms will be found along with the adults in most habitats. The habitat in which this species is most abundant is under coral boulders on the reef crest. However, specimens may be found in most other intertidal habitats as well as on the reef slope and extending down to the boundary with the off-reef floor. The specimens which are found sub-tidally are larger usually than those found intertidally. The average size of specimens varies from one reef zone to another, but on the reef crest it is about 47 mm arm radius. However, its size is dependent on the amount of autotomy which has occurred recently. The species can grow to about 90 mm. Some of the specimens encountered at the edge of the off-reef floor possess epiphytic ctenophores crawling over the arms of the starfish.

Echinaster stereosomus Fisher,1913

At Heron Reef, this species is found near the base of the reef slope. It occurs on the rocky reefs off southern Queensland more frequently than it does at Heron Reef.

Family Asteriidae

Coscinasterias calamaria (Gray, 1840)

This is primarily a southern species (Clark, 1946). Barrier Reef specimens are small, up to 30 mm arm radius, compared with the much larger individuals found on the mainland coast. This species is capable of asexual reproduction by binary fission. At Heron Reef, *C. calamaria* maintains small patches of moderate abundance by asexual reproduction. Indeed, it appears unlikely that specimens grow sufficiently large to become sexually mature at Heron Reef.

2.4 Discussion

The following species represent new records for Heron Reef:

Iconaster longimanus, Asteropsis carinifera, Dactylosaster cylindricus, Fromia elegans, Linckia multifora, Ophidiaster armatus, Ophidiaster lioderma, Ophidiaster robillardi, Tamaria megaloplax, Asterina anomala, Disasterina abnormalis, Tegulaster emburyi, Mithrodia clavigera, Echinaster stereosomus and *Coscinasterias calamaria*.

This study has provided the most southerly records from Great Barrier Reef waters of *Iconaster longimanus, Asteropsis carinifera, Dactylosaster cylindricus, Fromia elegans, Linckia multifora, Ophidiaster armatus, Ophidiaster lioderma, Ophidiaster robillardi, Tamaria megaloplax, Asterina anomala, Disasterina abnormalis, Mithrodia clavigera* and *Echinaster stereosomus*.

Single specimens of both *Ophidiaster lioderma* and *Tegulaster emburyi* were recorded at Heron Reef during this study and these represent the only known specimens of these species apart from their holotypes. Additionally, this study has provided the first record of the predominantly temperate species, *Coscinasterias calamaria* on the Great Barrier Reef. *Euretaster insignis*, which has not been recorded in the vicinity of a reef of the Great Barrier Reef, was found on the off-reef floor between Heron and Wistari Reefs.

Ophidiaster watsoni and *Anseropoda rosacea* were recorded from Heron Reef by Endean (1957) but were not located during this study. The taxonomic position of the former species is unclear. *Anseropoda rosacea* either is very uncommon at present, or primarily inhabits the sandy bottom of the lagoon which was not sampled extensively. *Ophidiaster hemprichi* and *Ophidiaster lorioli* occur at Heron Reef (Marsh pers. com.), but were not located during this study. *Halityle regularis* and *Andora popei* have been recorded from the off-reef floor in the vicinity of Heron Reef, by Baker and Marsh (1974) and Rowe (1977) respectively. These species were not located during this study as the off-reef floor was not sampled as intensively as were the shallow-water zones.

Because of its southerly position on the Great Barrier Reef, some predominantly sub-tropical asteroid species (e.g. *Ophidiaster confertus* and *Coscinasterias calamaria*) occur at Heron Reef but appear to not occur further north on the Great Barrier Reef. Additionally, some predominantly mainland species (Endean, 1957) occur either on, or in close proximity to, reefs of the Great Barrier Reef. The biogeographical study of Endean (1957) has shown that a distinction must be made between the asteroid species which occur predominantly on coral reefs of the Great Barrier Reef and those which occur elsewhere in Queensland waters. The results of the present study are in accord with this view. Clearly, there is a coral-reef asteroid fauna exemplified by that of Heron Reef, which is different from that of off-reef waters. However, as noted by Endean (1957), some species which occur on reefs of the Great Barrier Reef are not exclusively coral-reef species. For example, species such as *Archaster typicus*, *Protoreaster nodosus*, *Ophidiaster confertus*, *Tamaria megaloplax*, *Asterina nuda*, *Patiriella pseudoexigua*, *Anseropoda rosacea* and *Coscinasterias calamaria* occur predominantly in habitats other than those provided by coral reefs.

It seems likely, because of the extremely southern position of Heron Reef and other reefs in the Capricorn and Bunker Group, that many of the predominantly coral-reef species do not occur there with the same abundance as they do further north where physical conditions such as low water temperature on the reef flat in winter may be less extreme. Additionally, the relative isolation of this group of islands and reefs from the rest of the Great Barrier Reef might influence the abundance of those species with a low capacity for larval dispersal. However, these factors do not appear to affect the abundance of species that are common throughout the Great Barrier Reef. At higher latitudes, such as that of Heron Reef, the factors just mentioned might increase the abundance range between the most common and the rarest species. This would be reflected in the extent of sampling that would be required to locate most of the species that occur in the locality.

When current asteroid species lists for Heron Island and other reefs of the Capricorn Group are compared with those of recent studies of the North Pacific coral-reef Asteroidea (Yamaguchi, 1975 b; Marsh, 1977) it is apparent that some of the species that occur at Guam or Palau (e.g. *Archaster typicus*, *Celerina heffernani*, *Fromia indica*, *Fromia monilis*, *Nardoa tuberculata*, *Nardoa tumulosa*, *Neoferdina offreti*, *Asterina corallicola* and *Echinaster callosus*), have not been recorded from the Capricorn Group. On the other hand, some of the species that have been recorded from Heron Island and other reefs of the Capricorn Group (e.g. *Tosia queenslandensis*, *Iconaster longimanus*, *Fromia elegans*, *Nardoa pauciforis*, *Nardoa rosea*, *Neoferdina cumingi*, *Ophidiaster armatus*, *Ophidiaster lioderma*,

Disasterina abnormalis, *Disasterina leptalacantha* and *Tegulaster emburyi*), have not been recorded from either Guam or Palau.

Future investigations may reveal that some of the above similar but geographically separated coral-reef species (e.g. *Fromia indica* and *Fromia elegans*) are conspecific. However, future investigations may also confirm the restricted distributions of some of the species mentioned above.

Most of the coral-reef asteroids found on the Great Barrier Reef, including Heron Reef and other reefs of the Capricorn Group, have strong affinities with coral-reef asteroids of the Western Pacific region as noted by Endean (1957). However, a few species appear endemic to the reefs of the Capricorn Group or are essentially sub-tropical species that have extended their ranges to include the southernmost reefs of the Great Barrier Reef.

CHAPTER 3

HABITAT

3.1 Introduction

The spatial distribution of coral-reef starfish has been studied at several different scales. The physical, biological and historical parameters that explain the distribution of these species on the global scale (Clark and Rowe, 1971) will not directly explain the spatial pattern of an assemblage on a single reef. The small-scale distribution of *Linckia laevigata* on the fringing reef at Guam, and some factors which determine the abundance of this species in different habitats, were described by Strong (1975). The distribution and movements of *Linckia laevigata* at Lizard Island were examined by Thompson and Thompson (1982). Laxton (1974) suggested that *Linckia laevigata* may alter its distribution following outbreaks of *Acanthaster planci*. The distribution of an assemblage of starfish on a reef that is not known to have undergone outbreaks of *Acanthaster planci* has not been studied previously.

Feeding of starfish was extensively reviewed by Sloan (1980) and Jangoux (1982) and there have been many detailed examinations of the diets, competitive interactions and niche separation of colder water species (Blankley, 1984; Menge, 1972 a, 1972 b, 1981; Menge and Menge, 1974). The ecology of the tropical omnivorous, Atlantic species *Oreaster reticulatus* has been extensively studied by Scheibling (1980, 1981 a, b, 1982). The known diets and habitat preferences of the Indo-West Pacific coral-reef species have been tabled by Yamaguchi (1975 b). The contribution of these aspects of niche specialisation to the co-existence of many asteroid species in the coral reef ecosystem is poorly understood.

The general correlation between food supply and growth in asteroids has been discussed (Mead, 1900; Wolda, 1970; Paine, 1976). Species such as *Acanthaster planci* and *Culcita novaeguineae* are known to prey on hard corals (Endean, 1969; Yamaguchi, 1975 b; Glynn and Krupp, 1986). The response of *Acanthaster planci* to different prey species has been studied (Ormond, Hanscomb and Beach, 1976). *Asterina anomala, Asterina burtoni, Ophidiaster granifer* and *Gomophia egyptiaca* are known to feed on sponges and ascidians (Thomassin, 1976; Yamaguchi, 1975 b) while *Astropecten polyacanthus* and other members of its genus are known to prey on molluscs (Christensen, 1970; Ribi and Jost, 1978; Jost, 1979). *Coscinasterias calamaria* along with most members of the Order Forcipulatida, which are primarily inhabitants of temperate waters, are known to also prey on molluscs (Sloan, 1980; Jangoux, 1982).

However, the vast majority of coral-reef starfish are thought to be general detritovores and to feed primarily on the epibenthic felt (Thomassin, 1976) which is widely distributed throughout the reef environment. The possibility of ciliary nutrition (filter feeding) in some asteroids was raised by Gemmill (1915).

With the exception of what were once considered "primitive" genera such as *Astropecten* and *Luidia* (see Blake, 1987), starfish generally feed by everting their stomach over the substrate and digestion is external (Blake, 1990). However, the forcipulate species *Heliaster helianthus* is known to possess a flexible feeding habit involving both intra-oral and extra-oral feeding (Tokeshi, 1991). When *Acanthaster planci* and *Culcita novaeguineae* feed on hard coral they evert their stomachs and leave white feeding scars where the living tissue has been digested off the skeleton (Yamaguchi, 1975 b). Predominantly epibenthic feeders such as members of the genera *Linckia, Nardoa* and *Ophidiaster* leave no such feeding mark to indicate the position of their everted stomach when they are removed from the substrate while feeding (Yamaguchi, 1975 b).

While there have been extensive studies of the micro-habitat requirements of some coral-reef animals, for example gastropods (see e.g. Kohn and Leviten, 1976; Leviten and Kohn, 1980; Reichelt, 1982) there has been little work done on this aspect of coral-reef asteroid ecology. The role of habitat complexity in determining the population densities of species of gastropod within the coral reef ecosystem was discussed by Kohn (1968). However, the temporal scale of many community studies is often insufficient to determine the stability, or otherwise, of the observed community structure.

The degree to which species specialise in their use of habitat and other resources is a major component of the complexity of any assemblage (Klopfer, 1959; Klopfer and MacArthur, 1960, 1961; MacArthur and Levins, 1964, 1967; May and MacArthur, 1972. Reichelt (1982) regarded the availability of refuges from predation, desiccation and turbulence as determining factors in the spatial pattern of many intertidal species of gastropod. Kohn (1968) suggested that habitat complexity and the resultant spatial heterogeneity may directly determine the diversity of species assemblages.

Other authors have stated that the species assemblages that they examined were predominantly random. Guilds of species which use their habitat in a similar fashion were proposed by Sale (1976, 1977). In determining the spatial distribution of organisms, many authors propose the use of null (or neutral) models to prevent random events from being misinterpreted as meaningful biological pattern (Connor and Simberloff, 1979; McGuinness, 1984). However, caution over the misuse of inappropriate null models, which incorrectly reject real pattern, has been suggested by other authors (Dunbar, 1980; Quinn and Dunham, 1983; Roughgarden, 1983; Gilpin and Diamond, 1982).

3.2 Methods

At the completion of each traverse, all specimens were identified, counted and measured. The estimated area of the traverse was also recorded. While traverses were not stratified with respect to habitat, the zone of maximum density (primary habitat) was quite apparent for some species. Where a species occurred in more than one zone, the zones of less-frequent occurrence were referred to as secondary habitats. The distinction between primary and secondary habitat was much clearer in some species than in others. Some species were found so rarely that their range of habitat is unknown. In these cases, the zone in which they were located is regarded as the primary habitat.

25 specimens of *Linckia laevigata* were tagged with small plastic clothing tags which were inserted through the body wall of one arm, near its base, in such a way that a number inscribed on each tag was visible on close examination of the specimen. These starfish were then released at a site that provided no refuge other than under boulders. All boulders within a radius of 30 meters were overturned during attempts made to relocate them at 24 hour intervals. Twenty-five specimens of each of five additional species, *Linckia guildingii*, *Linckia multifora*, *Nardoa novaecaledoniae*, *Nardoa pauciforis* and *Echinaster luzonicus*, were released following similar tagging. Other methods of tagging, such as stains (Loosanoff, 1937; Feder, 1955; Vernon, 1937), were not successful because of low intensity of staining.

The feeding of starfish at Heron Reef was not examined in detail. When starfish specimens were collected, their stomachs were often found everted over the substrate. The approximate size of the stomach was noted along with the type of food material on which they appeared to be feeding. Any mark that their feeding activity may have left on the surface of the substrate was recorded.

3.3 Results

Culcita novaeguineae did not occur commonly in any habitat that was sampled by the shallow-water traverses. This species appeared to be more abundant in coral pools adjacent to the lagoon than on either the reef crest or reef flat at the western end of the reef. *Linckia guildingii*, *Linckia laevigata*, *Nardoa novaecaledoniae* and *Nardoa pauciforis* appeared to be slightly more abundant in one zone (primary habitat) than another (secondary habitat), but this difference was not clear, and was not quantified. All the remaining species occurred with either much greater densities in their primary habitats compared with their densities in secondary habitats, or were found in only one major zone.

The species that are regarded as exposed sometimes were found under boulders or coral rubble but no regular pattern of concealment was apparent in these species. The cryptic species varied in their type of refuge, which ranged from small rubble to large boulders.

The primary (1') habitat, secondary (2') habitat and general pattern of concealment of each species are listed in Table 3.1. These data are not quantitative but represent a general impression of the overall distribution pattern for each species. While some species are extremely restricted in their spatial distribution, others are widespread and it was not possible to estimate the abundance of each species, in each zone, in each sampling period when specimens were primarily required for size-frequency and reproductive analysis.

The general location of each species at Heron Reef and the known diet of each of the species (after Yamaguchi, 1975 b) are listed in Table 3.2. While diet was not studied in detail, observations made during this study confirm those of Yamaguchi that most species of starfish feed on epibenthic felt.

Table 3.1 The species of Asteroidea, primary habitat, secondary habitat, and their habit (excluding those species that occur predominantly in the off-reef floor zone).

SPECIES	1' HABITAT	2' HABITAT	HABIT
Astropecten polyacanthus	floor	flat	exposed
Iconaster longimanus	slope	floor	exposed
Culcita novaeguineae	flat	crest, slope	exposed
Acanthaster planci	slope	flat	both
Asteropsis carinifera	crest		cryptic
Dactylosaster cylindricus	crest		cryptic
Fromia elegans	slope	crest	exposed
Fromia milleporella	crest		exposed
Gomophia egyptiaca	slope	crest	both
Linckia guildingii	flat	crest	exposed
Linckia laevigata	flat	crest	exposed
Linckia multifora	crest	slope	both
Nardoa novaecaledoniae	crest	flat	exposed
Nardoa pauciforis	flat	crest	exposed
Nardoa rosea	floor	flat	exposed
Neoferdina cumingi	slope	crest	both
Ophidiaster armatus	floor	crest	exposed
Ophidiaster confertus	crest		cryptic
Ophidiaster granifer	crest		cryptic
Ophidiaster lioderma	crest		cryptic
Ophidiaster robillardi	crest		cryptic
Ophidiaster watsoni	crest		cryptic
Anseropoda rosacea	flat		cryptic
Asterina anomala	crest		cryptic
Asterina burtoni	crest		cryptic
Disasterina abnormalis	crest		cryptic
Disasterina leptalacantha	flat	crest	cryptic
Tegulaster emburyi	crest		cryptic
Mithrodia clavigera	crest		exposed
Echinaster luzonicus	crest	flat, slope	both
Coscinasterias calamaria	crest		cryptic

Table 3.2
Diet and location of each species (excluding those species that occur predominantly in the off-reef floor zone). In some species the diet is unknown.   asc = ascidian

SPECIES	DIET	LOCATION
Astropecten polyacanthus	mollusc	fine sand 0-30 m
Iconaster longimanus	felt	rubble 20-30 m
Culcita novaeguineae	coral, felt	sand, rubble 0-25 m
Acanthaster planci	coral	live coral 0-25 m
Asteropsis carinifera	-	sand under rock
Dactylosaster cylindricus	-	rock under rock
Fromia elegans	felt	rubble 0-25 m
Fromia milleporella	-	rubble
Gomophia egyptiaca	sponge, asc	sand, rubble 0-20 m
Linckia guildingii	felt	sand, rubble
Linckia laevigata	felt	sand, rubble
Linckia multifora	felt	attached under rock
Nardoa novaecaledoniae	felt	sand, rubble
Nardoa pauciforis	felt	sand, rubble
Nardoa rosea	-	sand, rubble 0-30 m
Neoferdina cumingi	-	sand, rubble 0-20 m
Ophidiaster armatus	-	sand, rubble 0-30 m
Ophidiaster confertus	-	rock under rock
Ophidiaster granifer	sponge, asc	attached under rock
Ophidiaster lioderma	-	rubble under rock
Ophidiaster robillardi	-	attached under rock
Ophidiaster watsoni	-	under boulder
Anseropoda rosacea	-	sand
Asterina anomala	sponge, asc	attached under rock
Asterina burtoni	sponge, asc	attached under rock
Disasterina abnormalis	felt	attached under rock
Disasterina leptalacantha	felt	attached under rock
Tegulaster emburyi	-	attached under rock
Mithrodia clavigera	-	rubble
Echinaster luzonicus	felt	sand, rubble 0-25 m
Coscinasterias calamaria	mollusc	attached under rock

In all species, excepting *Acanthaster planci*, the stomach was small, usually about the same area as the disc. At no time was the ingestion of large food material observed in any species but *Astropecten polyacanthus* and *Coscinasterias calamaria* were occasionally observed feeding on very small gastropods which were partially inside the mouth. In all other species the feeding was entirely extra-oral.

When the stomach was everted, the oral spines were oriented away from the mouth and both digestion and absorption occurred outside the body. At the completion of feeding, the stomach was withdrawn through the mouth and the oral spines were reoriented such that they occluded the mouth opening. All species commenced retraction of the stomach when removed from the substrate but the delay, before retraction was complete, varied from a few seconds in the case of epibenthic felt feeders with small stomachs to a few minutes in the case of *Acanthaster planci*. In all species, the oral spines could not reorient into the non-feeding (defensive) positions until the stomach was fully retracted inside the mouth.

The only two species which left feeding scars on the substrate were *Culcita novaeguineae* and *Acanthaster planci*. Both of these species feed on hard coral. *Culcita novaeguineae* was observed also feeding on bryozoan colonies at the base of the reef slope at a depth of 20 meters. All the other species had not altered the appearance of the substrate on which they had been feeding. *Fromia elegans*, *Linckia guildingii*, *Linckia laevigata*, *Linckia multifora*, *Nardoa novaecaledoniae*, *Nardoa pauciforis*, *Disasterina abnormalis*, *Disasterina leptalacantha* and *Echinaster luzonicus* were only observed to evert their stomach over substrate of either sand or coral-rock that was covered with a fine layer of organic material (epibenthic felt). *Ophidiaster granifer* and *Asterina burtoni* were observed with their stomachs everted over both solitary and colonial ascidians, but these two species also everted their stomachs over the epibenthic felt.

The results of movement studies of *Linckia laevigata* showed that this species is capable of moving at least 30 meters in a 24 hour period. Only 12 out of 25 specimens, which were released on coral rubble substrate at the outer, northern reef flat were able to be relocated 24 hours later. Eight specimens had not moved, three specimens had moved one meter, one specimen had moved 15 meters, but 13 specimens had moved a distance greater than 30 meters, and were not relocated. After a further 24 hours only two tagged specimens and no untagged specimens could be relocated in the vicinity of the point of release. Of the 25 specimens of six other species that were released following tagging, no tagged specimens were observed when attempts were made to relocate them after an interval of two months. It is possible that the small plastic tags were lost from the arms of these specimens.

3.4 Discussion

It can be seen from Table 3.1 that 16 of the 31 asteroid species recorded from Heron Reef were found in more than one of the major coral reef zones during this study. *Echinaster luzonicus* was found in all zones. The less common species were not encountered often enough for the data to show their complete distribution. With the exception of *Culcita novaeguineae*, *Linckia guildingii*, *Linckia laevigata*, *Nardoa novaecaledoniae*, *Nardoa pauciforis* and *Echinaster luzonicus*, the asteroids at or in the vicinity of Heron Reef can be divided into reef flat, reef crest, reef slope and off-reef floor species.

Species that occurred with their highest abundance on the reef flat are *Culcita novaeguineae*, *Linckia guildingii*, *Linckia laevigata*, *Nardoa pauciforis* and *Disasterina leptalacantha*. All species, excepting *D. leptalacantha*, are a large size when fully grown (large-bodied) and lie fully exposed in the daytime. However, they might complete their early development under boulders on the reef crest (Yamaguchi 1973 a, b).

Disasterina leptalacantha is a small cryptic species found occasionally on the reef crest, or more often, under slabs of beachrock on the innermost part of the reef flat.

Linckia multifora, *Nardoa novaecaledoniae* and *Echinaster luzonicus* occurred with highest abundance on the reef crest. Species which were found exclusively on the reef crest are *Asteropsis carinifera*, *Dactylosaster cylindricus*, *Fromia milleporella*, *Ophidiaster confertus*, *Ophidiaster granifer*, *Ophidiaster lioderma*, *Ophidiaster robillardi*, *Asterina anomala*, *Asterina burtoni*, *Disasterina abnormalis*, *Tegulaster emburyi*, *Mithrodia clavigera* and *Coscinasterias calamaria*.

Predominantly reef slope species are *Iconaster longimanus*, *Fromia elegans*, *Gomophia egyptiaca*, *Neoferdina cumingi* and *Acanthaster planci*. Their distribution seems more closely allied to living coral than is that of the species which inhabit the reef flat or reef crest.

The off-reef floor is not a true coral reef environment although numerous solitary corals and alcyonarians occur. The asteroid fauna of this environment is made up chiefly of the widespread sub-littoral species *Astropecten polyacanthus*, *Pentaceraster regulus* and *Euretaster insignis* in the deeper water (30-40 meters). *Nardoa rosea*, *Ophidiaster armatus*, *Tamaria megaloplax*, *Echinaster stereosomus* and large individuals of *Linckia multifora* and *Echinaster luzonicus* are found in the shallower water (20-30 meters) near the base of the reef slope.

Of the 25 starfish species found on Heron Reef itself, 17 were located only in intertidal regions and an additional three were found predominantly in intertidal regions. The other five species have been found intertidally, but occur predominantly in subtidal habitats.

Some cryptic reef crest species such as *Asteropsis carinifera*, *Linckia multifora*, *Disasterina abnormalis*, *Echinaster luzonicus* and *Coscinasterias calamaria* occur under rocks which possess a sparse development of epibiota. This would appear to be a short lived micro-habitat. Other cryptic reef crest species such as *Gomophia egyptiaca*, *Ophidiaster granifer*, *Ophidiaster robillardi*, *Asterina anomala* and *Asterina burtoni* are more frequently associated with encrustations of sponges and ascidians (Yamaguchi, 1975 b). *Ophidiaster confertus* was found either under or attached to the side of large boulders on the reef crest. Other species such as *Dactylosaster cylindricus*, *Neoferdina cumingi*, *Ophidiaster lioderma* and *Tegulaster emburyi* were not located in sufficient numbers to establish any pattern of occurrence. The reef crest species, that are regarded as cryptic during the day, did not move into exposed locations at night and species could remain cryptic during periods of activity. The reef crest, interstitial environment which is composed of highly fragmented coral, continually percolated by sea water, might provide refuges from desiccation and predation for some species but all attempts to locate asteroids in this micro-habitat were unsuccessful.

The epibenthic felt which covers large areas of intertidal coral reef habitat is composed primarily of protozoans, algae and bacteria (Thomassin, 1976). At Heron Reef and on other reefs, this appears to be a substantial resource. It would be of interest to know if competitive exclusion of one or more species can occur when population densities are higher than those recorded at Heron Reef. The role of species specific enzymes in the digestion of different organisms composing the epibenthic felt has not been investigated to date. Each species might possesses enzymes which facilitate the efficient exploitation of a different component of the epibenthic felt. A biochemical study of the gastric mucosa of each asteroid species would be needed to establish resource partitioning on this microscopic scale.

The feeding and movements of *Acanthaster planci* was studied by Keesing and Lucas (1992). This species has been shown to possess an enzyme which can efficiently digest the wax ester, Cetyl Palmitate, which is stored in the soft tissues of hard corals (Benson et al., 1975; Brahimi-Horn, 1989). Consequently, this asteroid must be regarded as a highly specialised predator of scleractinians. It is possible that other highly specialised enzymes occur in coral-reef asteroids. To exploit epibenthic felt efficiently, a scavenger would utilise mechanical or bacterial breakdown of algal cell walls and in other taxa this would be accomplished within a gut and the freed inter-cellular material would be subsequently assimilated by the organism. The epibenthic feeding asteroids evert their stomach over the felt and digestion occurs externally. These asteroids rely on enzymes to complete the digestion process but there are few known eucaryote enzymes which are capable of chemically digesting cellulose walls of algae. Diatoms, which are common in the epibenthic felt, have siliceous walls and the chemical digestion of this material, by any organism, would seem impossible. It is apparent, however, that many species of asteroid coexist on this same resource and further work is needed on possible dietary (enzyme) specialisation.

Predation on molluscs has been recorded in *Astropecten polyacanthus* and *Coscinasterias calamaria*. The paucity of mollusc-feeding coral-reef asteroids on Heron Reef is noticeable, compared with their high abundance in temperate waters (see Menge, 1975; Kwon and Cho, 1986; Nojima et al., 1986). Individuals of *Coscinasterias calamaria* found in the reef crest habitats were only small ($R < 20$ mm), as were the gastropod prey on which they were observed feeding. The mobility of starfish might be insufficient to allow large-scale foraging on a coral reef as well as sufficient aggregation for successful reproduction.

Blake (1983) suggested that the general body plan of molluscivorous starfish leaves them vulnerable to predators. The observed delay in oral spine relocation following the commencement of stomach retraction, together with the defensive nature of these spines, suggests that the oral region of a starfish may be especially vulnerable to predatory attack particularly during and immediately following starfish feeding. The absence of adult specimens of *Coscinasterias calamaria* and other molluscivorous species of starfish may be related to the dangers inherent in this type of feeding.

In summary, the coral-reef starfish that occurred at Heron Reef showed some inter-specific variation with respect to diet, but many species appeared to feed on the same food (epibenthic felt). These species also showed some inter-specific variation with respect to habitat, but in every coral reef zone, some species sought no refuge and occurred with exposed habits. Of the species that occur predominantly on coral reefs, clear examples of niche (dietary or microhabitat) specialisation are known only for *Culcita novaeguineae* and the predominantly subtidal species *Acanthaster planci*. Clear examples of competitive interactions were not observed during this study.

CHAPTER 4

POPULATION DENSITY

4.1 Introduction

It is well known that the scale of observation is critical for the determination of the spatial distribution pattern of a species. Differing scales of analysis can produce apparently differing results even with the same data. The properties or parameters that emerge from studies of communities can be dependent on which scale of organisation, space or time is chosen (Bradbury and Reichelt, 1982).

While the abundances of the various species of starfish will be partially determined by the small-scale distribution of scattered resources, the overall spatial distribution of each species will be a composite pattern influenced by food, refuge and predator abundance as well as aggregation behaviour (Patton *et al.*, 1991; Stevenson, 1992; Iwasaki, 1993). Each of these factors can vary at a number of scales.

For each species within this assemblage, population aggregation may vary either spatially (from one location to another) or temporally (over time at any one location). If there is an equal probability of locating a species at every point within its spatial distribution, then individuals of that species are distributed at random. However, if the geographical range of a species or its abundance variation within that range is attributed to either physical or biological parameters, then non-randomness of the spatial distribution of that species is directly implied.

If the scale of observation is such that individuals of a particular species would be expected to be distributed randomly throughout habitats, which themselves are distributed randomly in space, then the expected distribution of individuals in space will be clumped, not random. If low density populations of starfish are not expected to be distributed randomly, then density estimates can seriously underestimate the standard error of the mean. Failure to determine the degree of positive skewness in the density distribution results in poor repeatability. Population density estimates of non-random species are credible only when the extent of the positive tail of this distribution has been determined adequately.

4.2 Methods

Specimens were collected primarily for size-frequency and reproductive analysis. For logistical reasons, it was not possible to estimate the density of each species, in each zone, in each sampling period. For each species, the density on each traverse was calculated by dividing the number of individuals by the estimated area of the traverse. The mean density of each species was then calculated by taking the arithmetic mean of the 72 traverse densities. It is represented as the average number of individuals found per hectare.

Because starfish are not distributed randomly, the total number of individuals of each species divided by the total area is not equal to the mean of the individual traverse densities. The standard deviation of density was calculated from the 72 traverse densities and represents the overall variation in density across all the traverses.

Because of the nature of traverse sampling, the density of most species is only approximate. Exposed species are reasonably well estimated but the cryptic species are greatly underestimated in their abundance because not all coral rocks and boulders in each traverse were overturned. Although the undersurface of rocks was examined closely, the nature of the substrate would make detection of the smaller species less reliable than the detection of larger species. When specimens were located within the sediment under rocks, individuals that were buried deeply within the rubble or sediment under these rocks would not have been found.

It should be noted that, in addition to patchiness, the number of individuals of each species recorded on different traverses varied because of variation in the size of traverse. The total number of each species also varied between sampling periods as a result of variation in the number of traverses undertaken in each sampling period.

Disasterina abnormalis was sampled in detail because it occurred in one region at a high density. This was the only species that could be sampled in this manner and this species appeared to occur at this density in only one region. The mean individual density per square meter, over a number of contiguous quadrats, and a Chi-square value (with Yates' correction) of the inter-quadrat variation was calculated for *Disasterina abnormalis*. Twenty (metre square) quadrats were laid at Site 1 in April 1980 and again in July 1980. Forty (metre square) quadrats were laid at Site 2 in April 1980. All specimens occurring within the quadrats were counted and measured. It is to be noted that this was a region of northern reef crest where the density of this particular species was known from traverse data to be high.

4.3 Results

The data presented in Table 4.1 show the densities of all the intertidal asteroid species that occurred within traverses during this study at Heron Reef. For all species, the standard deviation was greater than the mean density. This, together with Figures 4.2 to 4.13, indicates variations in density that are greater than the expected Poisson variation. The results of quadrat density analysis of *Disasterina abnormalis* are shown in Table 4.2. The variation was analysed using chi square and individuals were clumped at the metre square scale.

Figure 4.1a graphs the linear relation between the total number of individuals and the total sample area. Figure 4.1b graphs the number of species in each of five (log) average density ranges. This illustrates how the average density of starfish species is distributed within this assemblage.

Figures 4.2 to 4.13 graph the population distribution of each of the common species, over the 72 traverses. Each graph displays the number of traverses on which a species occurred at a particular density. The density axis has been logged to facilitate the display of an extremely wide range of density.

Figures 4.14 and 4.15 are composite graphs of the population distributions of these species. Figure 4.14 graphs the population distributions of the six relatively abundant species, namely *Echinaster luzonicus*, *Disasterina abnormalis*, *Asterina burtoni*, *Nardoa novaecaledoniae*, *Linckia multifora* and *Linckia laevigata*. Figure 4.15 graphs the population distributions of the six less abundant species, namely *Asterina anomala*, *Ophidiaster granifer*, *Nardoa pauciforis*, *Linckia guildingii*, *Culcita novaeguineae* and *Fromia elegans*. The abundances of the 12 remaining species that occurred on traverses were very low and were not analysed.

Table 4.1

The density of each species that occurred on intertidal traverses expressed as mean density (number per hectare), standard deviation (S.D.) and number (N) of individuals.

SPECIES	MEAN DENSITY	S.D.	N
Culcita novaeguineae	0.14	0.48	15
Asteropsis carinifera	0.02	0.12	3
Dactylosaster cylindricus	0.01	0.12	1
Fromia elegans	0.10	0.42	16
Fromia milleporella	0.002	0.01	1
Gomophia egyptiaca	0.12	0.57	6
Linckia guildingii	1.27	2.63	116
Linckia laevigata	4.01	4.87	509
Linckia multifora	7.51	17.30	522
Nardoa novaecaledoniae	3.19	3.72	326
Nardoa pauciforis	1.60	1.77	187
Nardoa rosea	0.002	0.01	1
Ophidiaster armatus	0.02	0.11	4
Ophidiaster confertus	0.03	0.15	4
Ophidiaster granifer	1.56	2.67	116
Ophidiaster lioderma	0.02	0.15	1
Ophidiaster robillardi	0.58	2.58	24
Asterina anomala	0.23	0.58	17
Asterina burtoni	3.27	6.99	208
Disasterina abnormalis	5.68	10.04	500
Disasterina leptalacantha	0.23	1.31	7
Tegulaster emburyi	0.01	0.07	1
Echinaster luzonicus	16.16	24.67	1402
Coscinasterias calamaria	0.11	0.65	7

Figure 4.1a
Relation between total number of individuals and sample area

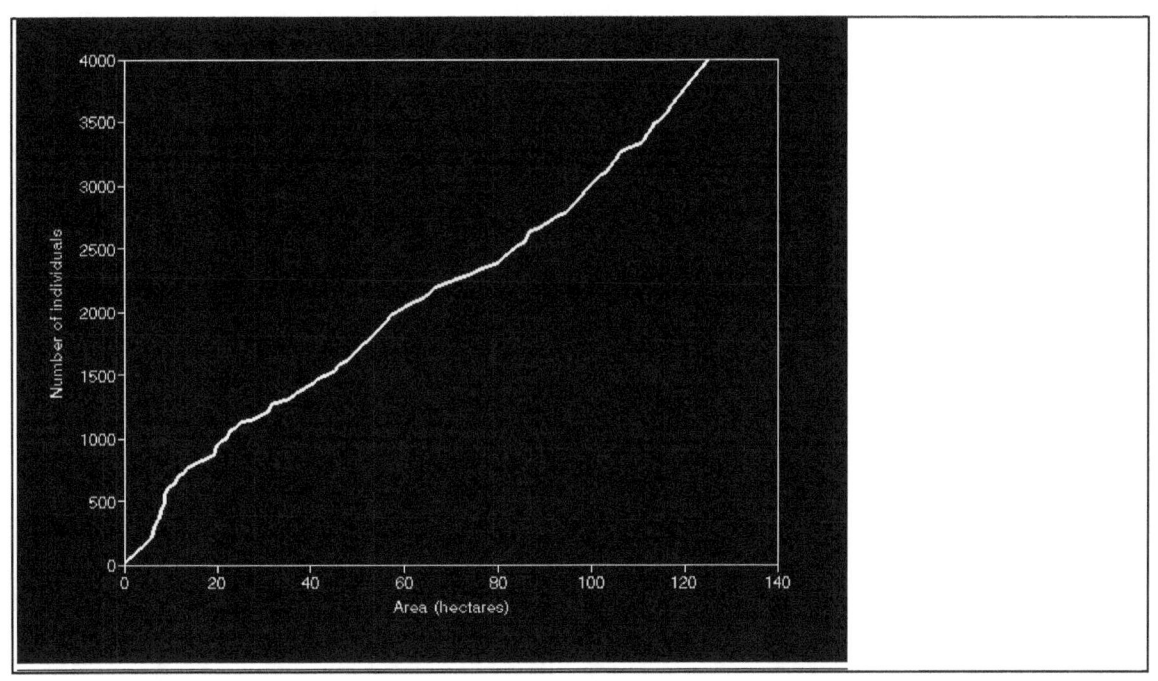

Figure 4.1b
The number of species in each of 5 (log) density categories

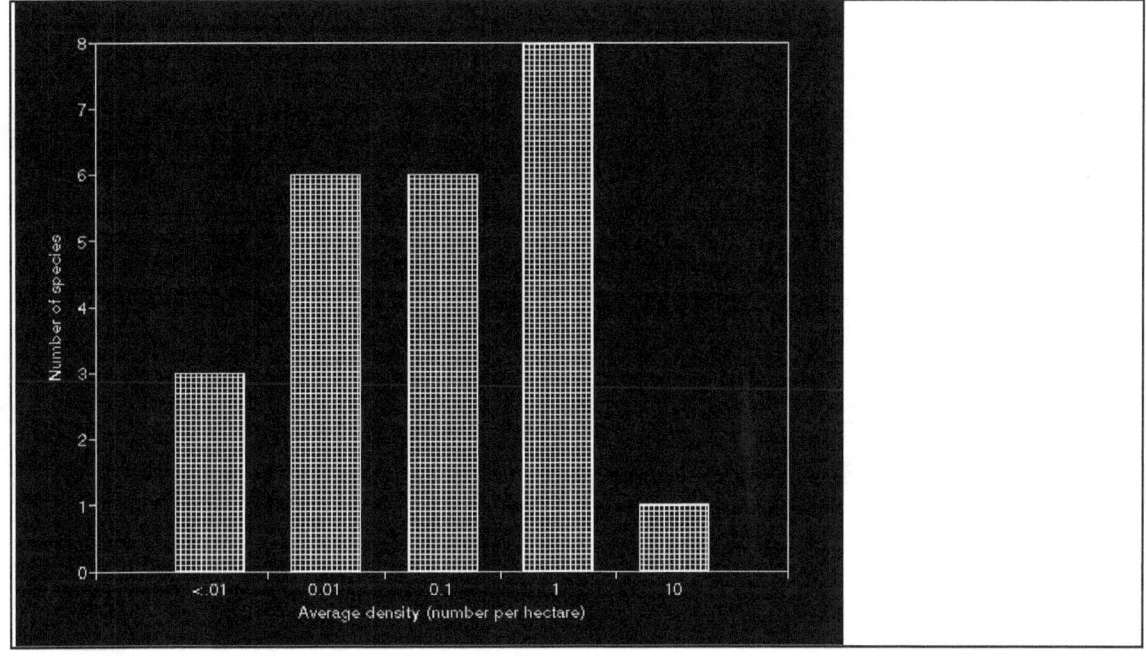

Table 4.2
Density and patchiness of *Disasterina abnormalis*.

The variation in number of individuals within adjacent square metre quadrats at two study sites and two sampling periods. DENSITY (the number of individuals per square metre), CHI-SQUARE (calculated from the inter-quadrat variation), PROB (the probability of this variation being random) and the NUMBER of individuals in the sample are tabled.

PERIOD		DENSITY	CHI-SQUARE	PROB.	NUMBER
APRIL 1980	SITE 1	8.4	58 (d.f.=24)	<.001	161
	SITE 2	0.7		N/S	29
JULY 1980	SITE 1	8.9	11 (d.f.=11)	N/S	98

Figure 4.2
Population distribution of *Culcita novaeguineae*.

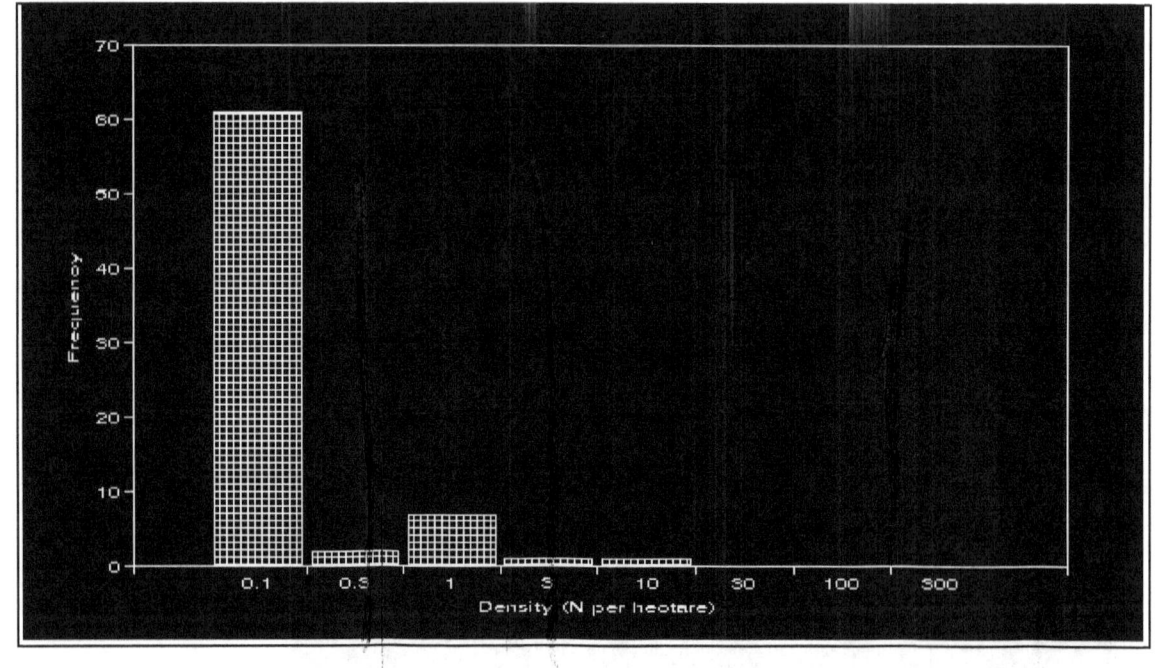

Figure 4.3
Population distribution of *Fromia elegans*.

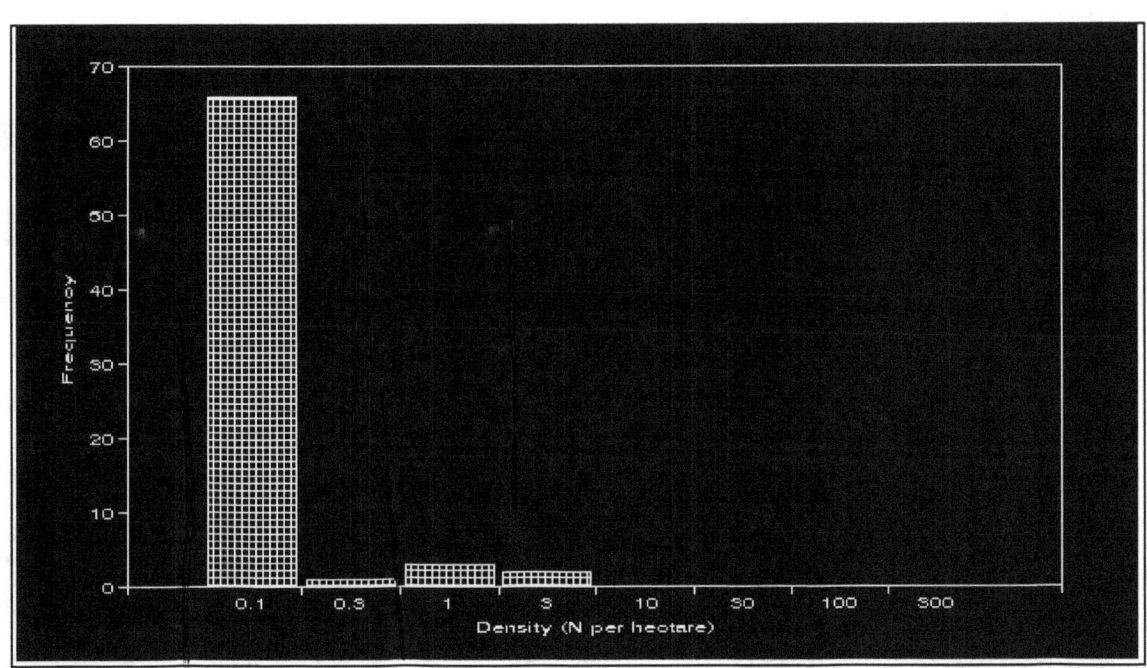

Figure 4.4
Population distribution of *Linckia guildingii*.

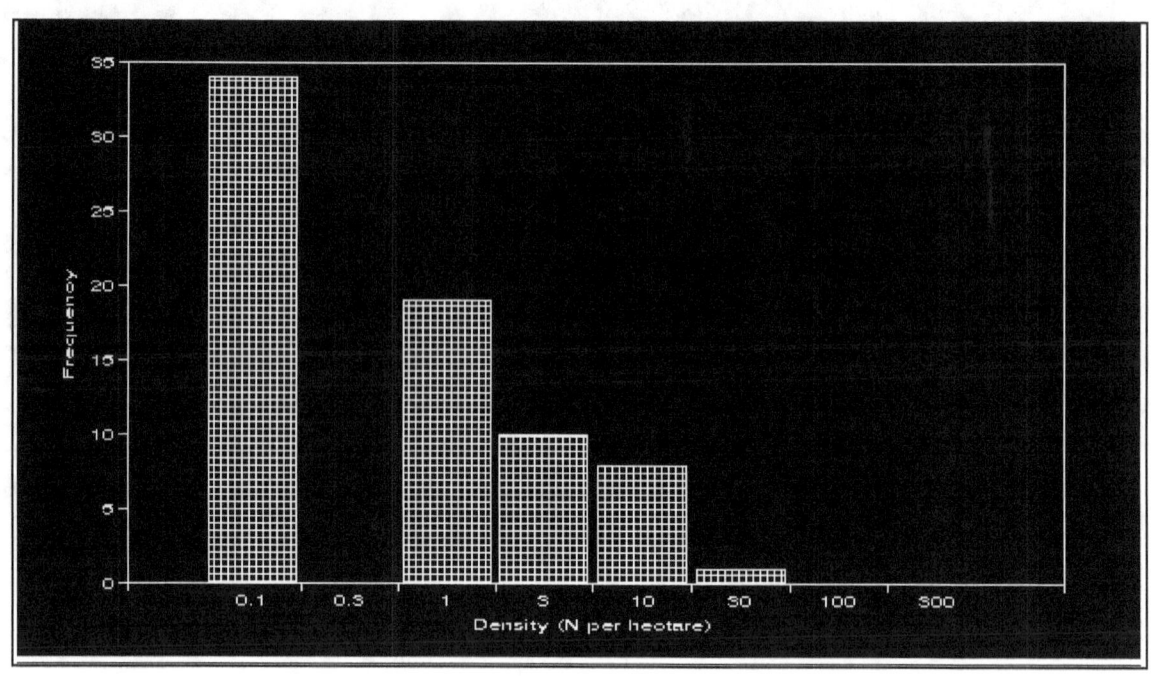

Figure 4.5
Population distribution of *Linckia laevigata*

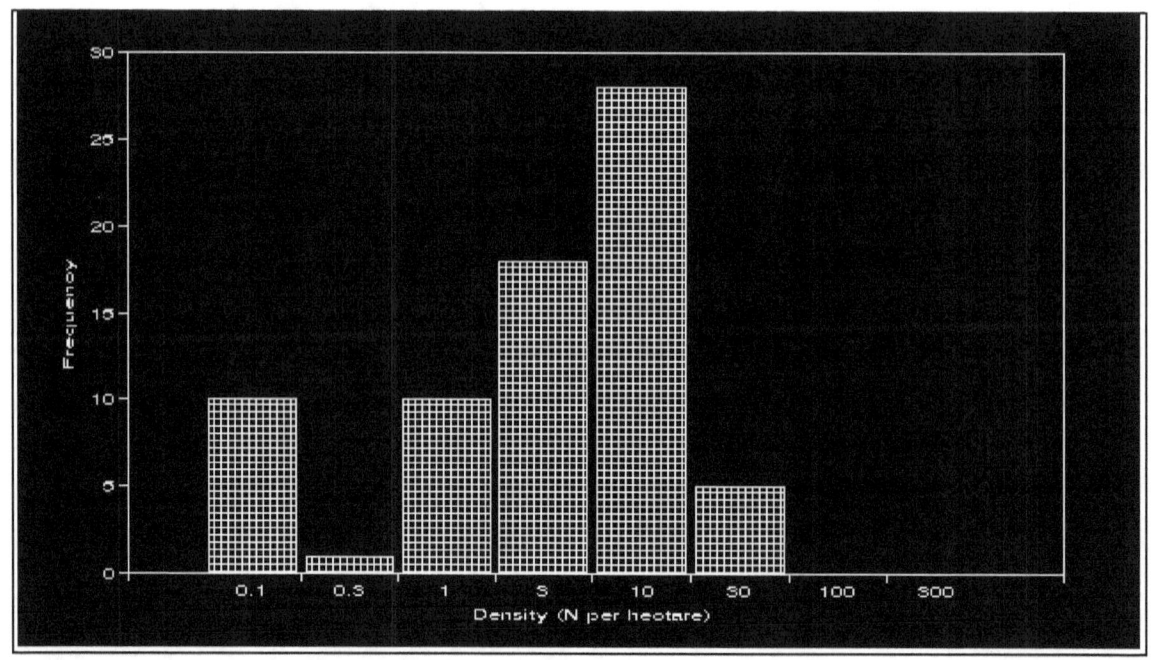

Figure 4.6
Population distribution of *Linckia multifora*

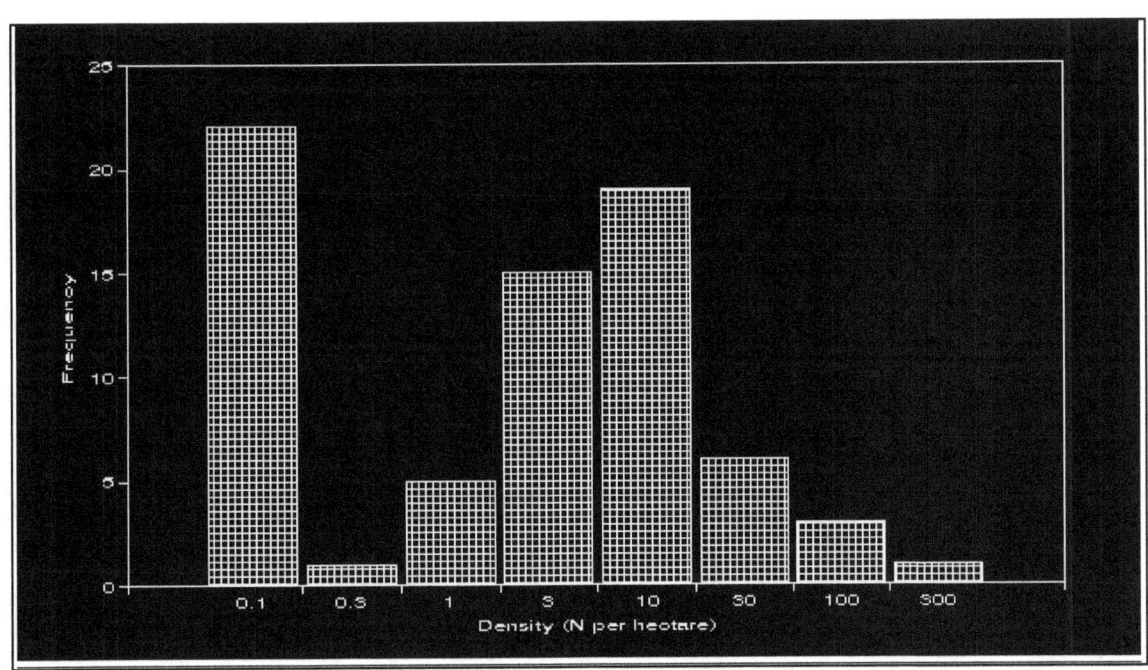

Figure 4.7
Population distribution of *Nardoa novaecaledoniae*

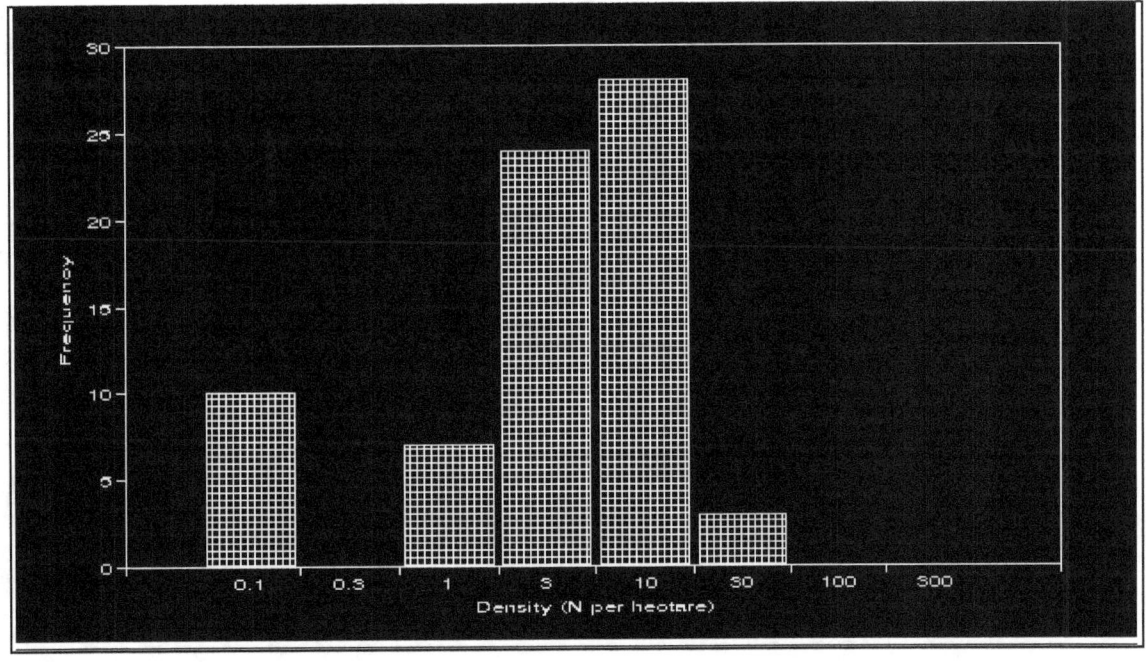

Figure 4.8
Population distribution of *Nardoa pauciforis*

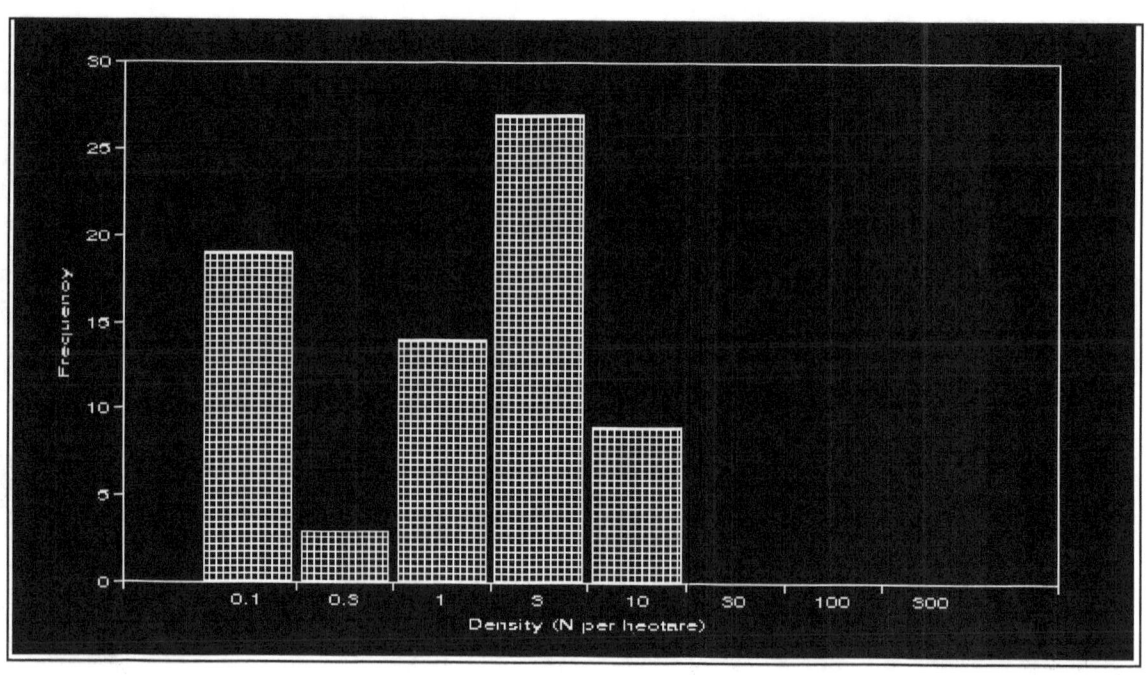

Figure 4.9
Population distribution of *Ophidiaster granifer*

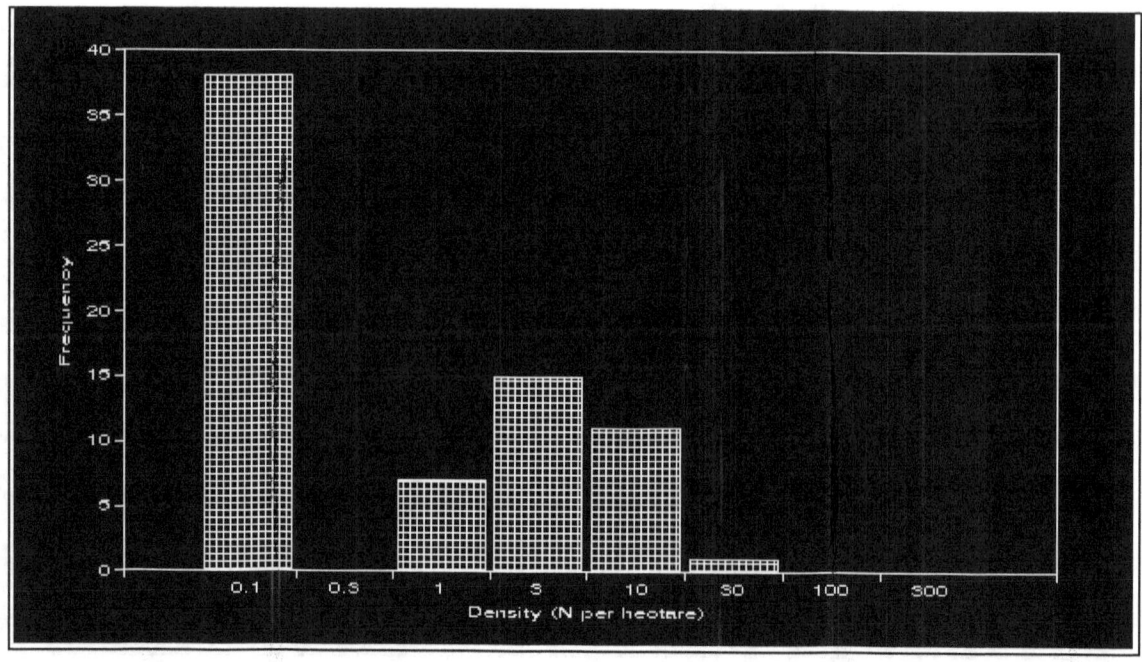

Figure 4.10
Population distribution of *Asterina anomala*

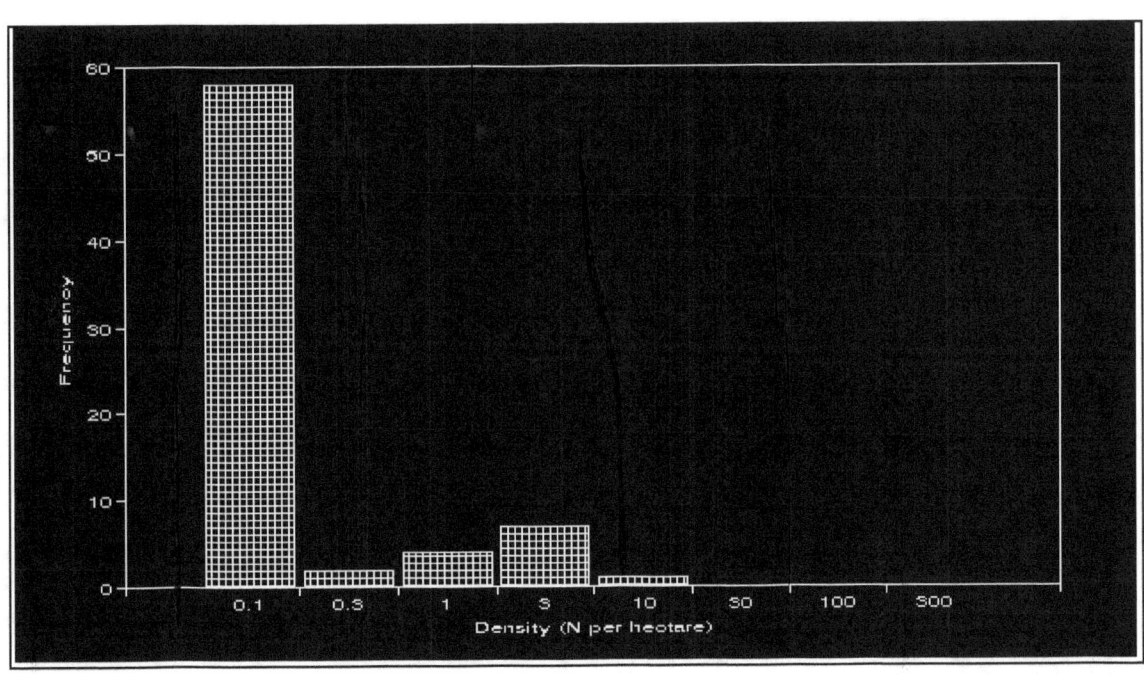

Figure 4.11
Population distribution of *Asterina burtoni*

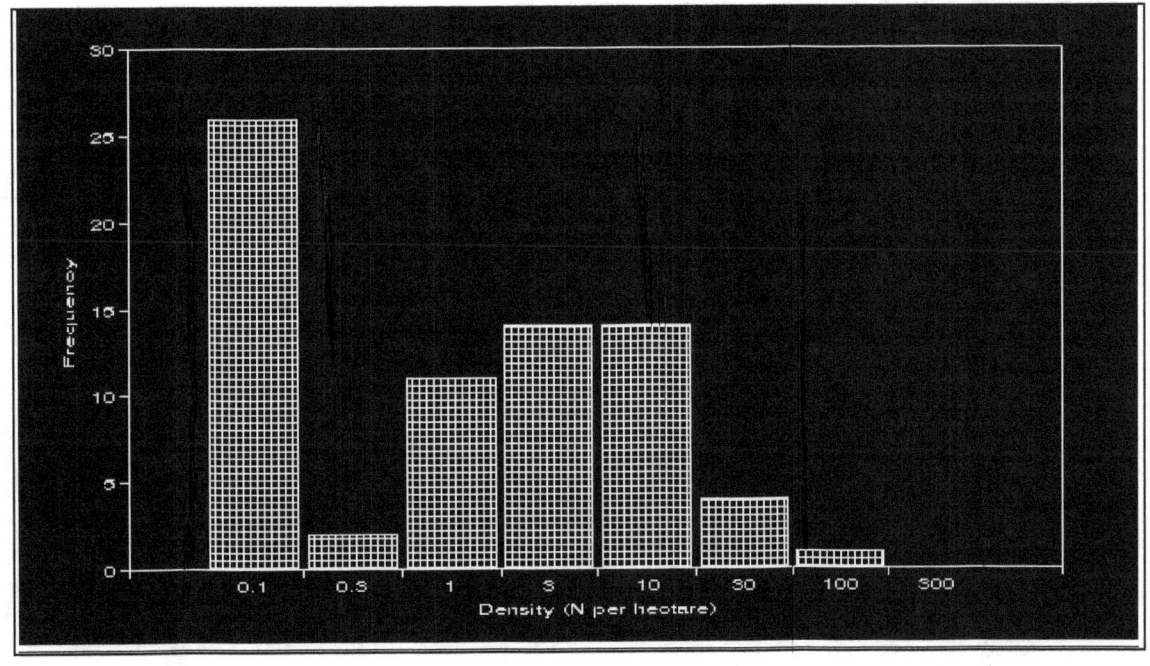

Figure 4.12
Population distribution of *Disasterina abnormalis*

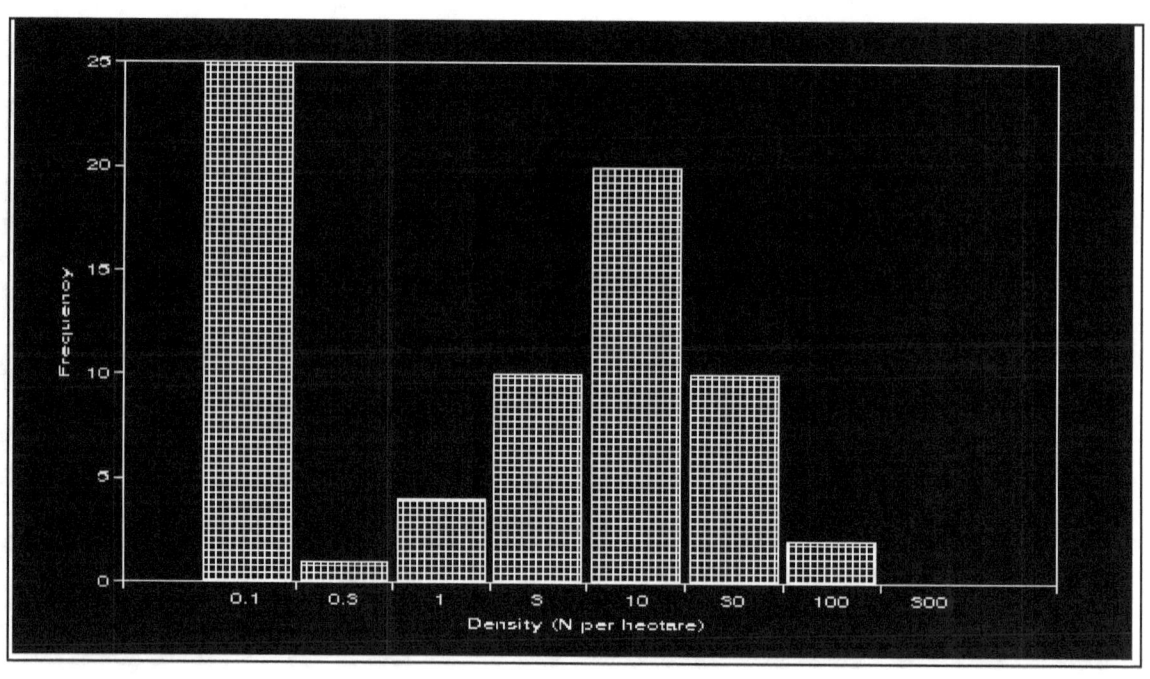

Figure 4.13
Population distribution of *Echinaster luzonicus*

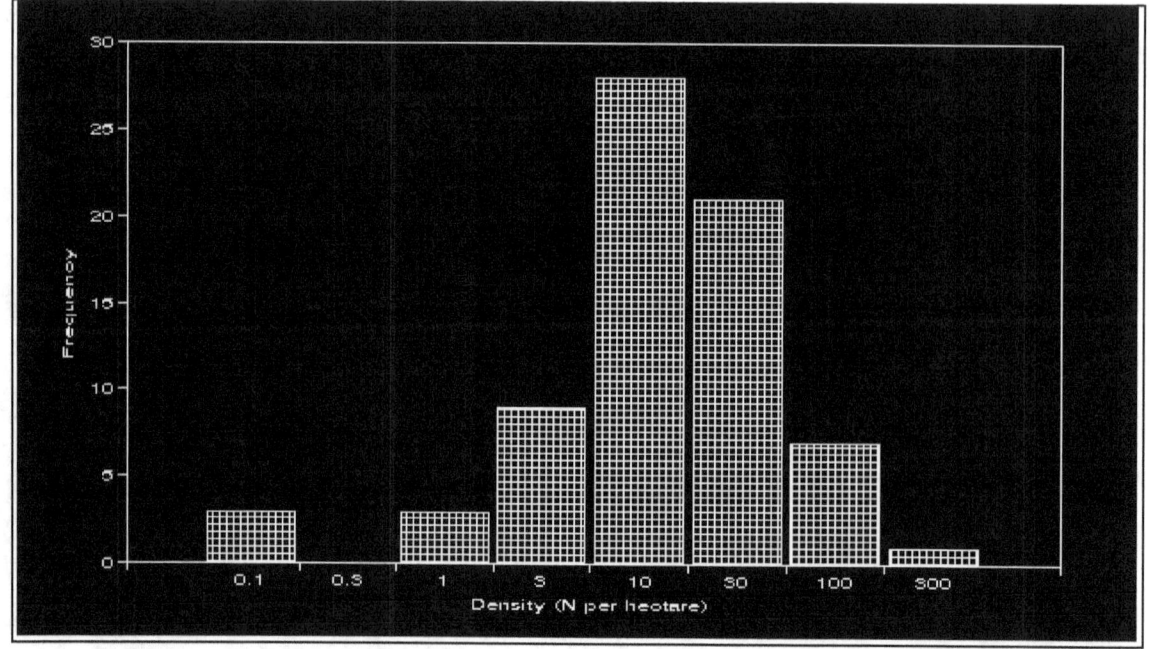

Figure 4.14

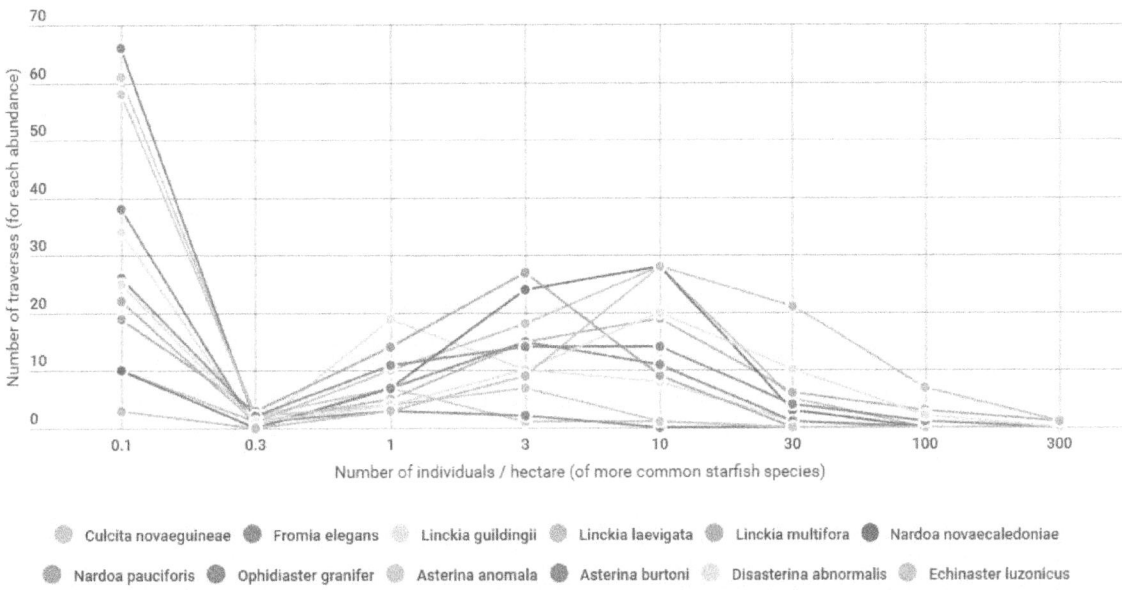

From 1978 to 1984 on Heron and Wistari Reefs, most species of starfish, including Acanthaster were rare. It is suggested that the observed minimum at 0.3 individuals / hectare would be expected to increase for each starfish species if predator abundance was reduced #CharoniaResearch

Figure 4.15

4.4 Discussion

The traverse data do not allow for a statistically valid comparison of density among different sites or different sampling periods. Four species of starfish appeared to demonstrate changes in density during the study period. Two of these species were capable of asexual reproduction and these species demonstrated periods of autotomy followed by periods of growth. These species were *Linckia multifora* and *Echinaster luzonicus*. Asexual reproduction following a sexual recruitment was suggested by Ottesen and Lucas (1982) and Yamaguchi and Lucas (1984) as the reason for the greatly different abundances of all asexually reproducing species at different places on the same reef.

The other two species that showed a large change in abundance were *Disasterina abnormalis* and *Asterina burtoni*. At the commencement of the sampling program, the density of *Asterina burtoni* appeared to be about half that of *Disasterina abnormalis* under boulders on the reef crest. The abundance change in *Asterina burtoni* could not be analysed accurately because *Asterina burtoni* did not occur in great abundance in any known habitat. As a result, it was not possible to sample its density using meter square quadrats. A temporal variation in the abundance of *A. burtoni* was recorded by Price (1981) in the Arabian Gulf. *Disasterina abnormalis* had periodic high recruitment with resultant changes in both its abundance and size-frequency distribution. For *Linckia multifora*, *Echinaster luzonicus* and *Disasterina abnormalis*, the changes in mean individual size results from periods of high recruitment which are discussed in the chapter on Population Stability. The range in abundance of *Acanthaster planci* in both outbreaking and non-outbreaking populations was investigated by Moran and De'ath (1992 a).

The results of quadrat sampling in an area of reef crest north of Heron cay where the density of *Disasterina abnormalis* was known to be high (Table 4.2, Site 1) showed an average density of 8.4 individuals per square metre. This region was the innermost part of the reef crest and was sheltered partially from heavy wave action by a bank of rubble which extended for about one kilometre. One hundred metres west of this rubble bank (Table 4.2, Site 2), in an otherwise similar region of the inner reef crest, the density of this species was less than one individual per square metre.

The number of individuals of *Disasterina abnormalis* per square metre showed too much variation in the April 1980 (Site 1) sample for the individuals to be randomly distributed at the time of sampling. However, in the July 1980 (Site 1) sample, the individuals were not significantly clumped. It is of interest that the density per square metre of this species did not differ significantly between these two sampling periods. The only difference was in the degree of aggregation. Antonelli and Kazarinoff (1988) regarded the degree of aggregation of *A. planci* as an important factor in the modelling of population regulation by predators.

The quadrat samples produced only a minute subset of the known number of species, because such a small area was sampled. It was not feasible to sample extensively by quadrat as the patchy distribution of all these species required a large-scale estimate of spatial density variation.

Environmental heterogeneity might account for the observed clumping of individuals that were found primarily under boulders or rubble, but does not explain the variation in abundance of exposed asteroids on traverses which crossed what appeared to be similar habitat. The effect of variation in physical parameters, such as depth of water, amount of siltation of substrate, and strength of wave action is unknown, and factors such as these might account for some of the observed differences in abundance.

It can be seen from Figures 4.2 to 4.13 that individuals of each of the species were usually either absent or reasonably well represented on traverses. Individuals of the more abundant species did not occur at higher densities on every traverse, but were located on more of the traverses and occurred more often at moderate and higher densities. Individuals of the less abundant species were absent from most of the traverses and occurred less often at the moderate densities and never at higher densities. The possibility of variation in the abundance of species from reef to reef also exists. This would be more noticeable if reefs maintain semi-closed circulations. The numbers of one species may gradually build up by local recruitment if larvae recruit to the parent population.

Some of the rarer species of coral-reef starfish are known only from their holotype or perhaps one or two paratypes and appear to exist at population densities which defy our normal understanding of population dynamics and reproductive strategies. It is not clear how these species survive and which, if any, ecological requirements or constraints limit their distribution or abundance. It is not known whether these species are rare because their necessary ecological requirements are met at only a small number of points or whether their rarity is a result of intense predation.

Recruitment involving survival to reproduction must occur at some points within the distribution of each species unless we are observing the process of extinction. Considering both the number of species involved and the fact that species such as *Tosia queenslandensis*, *Ophidiaster lioderma* and *Tegulaster emburyi* are considered rare throughout their geographical range, rare species must demonstrate physical or behavioural attributes which are adaptations to existence in low density populations. Levins and Culver (1971) suggested that specialised rare species might play a key role in ecosystem modulation and they raised the possibility of specialised predation or competition among rare species.

Any assumption of a species abundance indicating its successfulness or adaptive nature should be questioned and the concept of species adapted to live in sparse populations offered as partial explanation of the high diversity in many ecosystems. The influence of specific predation can result in the rarity of a species and adaptations to this might represent a viable survival strategy (Connell, 1970). Spawning aggregations, extended gamete survival and high gamete specificity, hermaphroditism, parthenogenesis and asexual reproduction are all ways of ensuring continuity of offspring in rare species. Certain very specialised species might occur only at a certain resource optimum and their populations will be limited to the number of these sites of optimum habitat.

It is not known to what extent population fluctuations are normal on coral reefs. In the larger species of starfish at Heron Reef, the overall impression was that population fluctuations were low compared with the fluctuations that are known to occur on other reefs and in temperate ecosystems.

Because the abundance data resulting from the traverse samples was biased towards large, exposed individuals, it would be unwise to use this traverse data for a direct density comparison over repeated sampling periods. The non-randomness of the spatial distributions of these starfish populations, as evidenced by the large range in local density that was recorded in the populations of many of the species, further limits the validity of such a density comparison. For this reason, the variation in mean individual size was considered a more appropriate measure of change in the population structure of the species.

CHAPTER 5

THE POPULATION SIZE STRUCTURE OF THE COMMONER SPECIES

5.1 Introduction

Many workers who have studied coral-reef asteroid populations have noted the adult-dominated size structure of these populations (Clark, 1921; Ebert, 1972; Yamaguchi, 1977 a). Similar findings have been made for asteroid populations from other communities (e.g. Paine, 1976). Although juvenile asteroids have been encountered on coral reefs, their abundance was so low and their appearance sufficiently different from that of the adults in some species, that juveniles have been placed in a species different from the adult (see Yamaguchi, 1975 b).

The population structure of some of the more common, large-bodied species of coral-reef asteroids such as *Linckia laevigata* has been studied (Yamaguchi, 1977 a; Laxton, 1974; Thompson and Thompson, 1982). In these studies, the size distributions of the asteroid were unimodal indicating either a large overlapping of generations or a dominant year class. If the latter alternative is true, the variation in growth rates within the population must be extraordinary to produce the observed size range. Most of the studies on *Acanthaster planci* occurring under non-outbreak conditions have shown a primarily adult population with small juveniles occurring only occasionally (Yamaguchi, 1973 a; Zann *et al*, 1987). In studies of large-bodied coral-reef asteroids, few individuals were found which were smaller than half the average size (Yamaguchi, 1973 a, 1973 b; 1977 a). In this chapter, the maximum size attained and the population structure of each of the commoner species constituting the asteroid fauna of Heron Island Reef will be investigated.

5.2 Methods

Specimens collected on traverses were allowed to resume an extended shape in a plastic bucket and were then measured using a plastic ruler. When animals were measured both the length from the mouth to the tip of the longest arm, and the average length from the mouth to the interradius were recorded to an accuracy of one millimetre. These are called major radius (R) and minor radius (r) respectively and are expressed in millimetres (mm). "R" was always measured along the ambulacral groove. After measurement, individuals were placed along with conspecifics in glass aquaria that were provided with fresh running sea water if they were needed for later experiments relating to their reproduction.

The major radius (R) is used as a measurement of overall size. The ratio of major radius (R) to minor radius (r) is referred to as "R/r" or "R:r". Because it is a ratio it has no units. It is a measurement of the degree of arm elongation and is of taxonomic significance.

In addition to specimens collected on traverses at Heron Reef many sub-tidal specimens of *Fromia elegans* and specimens of *Disasterina abnormalis* found during the quadrat study were measured. This is the reason for the difference in sample size between the tables of abundance and mean size.

Juvenile asteroids of most common species were located under boulders on the reef crest. Their identification, although initially difficult, was always possible following microscopic examination and reference to earlier studies (Clark, 1921; Yamaguchi, 1975 a, 1975 b, 1977 a).

5.3 Results

Tables 5.1 - 5.10 summarise the mean size data of all individuals recorded in each sampling period for each of the common species. ANOVA tables showing the significance of variations of major radius (R) are included. Figures 5.1a - 5.10a graph the frequency distribution of major radius (R). Figures 5.1b - 5.10b graph the frequency distribution of major radius / minor radius (R/r). Figures 5.1c - 5.10c graph the relation between minor radius (r) and major radius (R). Figures 5.1d - 5.10d graph the relation between major radius / minor radius (R/r) and major radius (R). The relation between these two radii is variable, but it is frequently used as a taxonomic distinction. Minor radius (r) was not measured in August and November 1978 and these data are excluded from the ANOVA. Table 5.11 is a summary of the size data for each of the 24 species that occurred on intertidal traverses.

Juveniles of *Fromia elegans*, *Linckia laevigata*, *Nardoa novaecaledoniae* and *Nardoa pauciforis* occurred rarely. The size-frequency distributions of major arm radius (R) show clearly the paucity of small individuals (R less than half the mean R) in the populations of these species. *Linckia guildingii*, *Linckia multifora*, *Ophidiaster robillardi*, *Asterina anomala*, *Echinaster luzonicus* and *Coscinasterias calamaria* reproduce asexually. Small individuals of these species, resulting from autotomy or fission, occurred intermittently throughout the period of the study . Juveniles of *Asterina burtoni* and *Disasterina abnormalis* occurred throughout the period of the study. The size- frequency distributions of major arm radius (R) show the abundance of small individuals in these species. *Ophidiaster granifer* was not common, but its highly skewed size-frequency distribution shows the presence of medium sized individuals (R half the mean R). With the exception of one specimen each of *Culcita novaeguineae* (R=50 mm) and *Gomophia egyptiaca* (R=10 mm), juveniles of less common species did not occur.

Table 5.1 *Fromia elegans*

The mean (R mm and R/r), standard deviation (R mm and R/r) and sample size (N) for each sampling period and a grand mean of both R (mm) and R/r, and total sample size are tabled.

PERIOD	MEAN(R)	S.D.	MEAN(R/r)	S.D.	N
JULY 1981	29.29	6.22	4.48	0.48	24
JANUARY 1982	32.65	3.11	3.74	0.27	34
MAY 1982	31.01	4.93	3.87	0.40	76
OCTOBER 1982	34.48	4.15	3.97	0.26	29
DECEMBER 1982	31.95	4.44	3.90	0.32	20
TOTAL	31.74		3.94		183

*********** ANOVA **********

Variation in mean Major Radius (R) with respect to time.

Source of Variation	df	SS	MS	F
Among	4.0	431.0	107.8	4.9
Within	178.0	3885.9	21.8	
Total	182.0	4316.9		

$p < .01$

Figure 5.1a *Fromia elegans*
Frequency distribution of major radius (R mm)

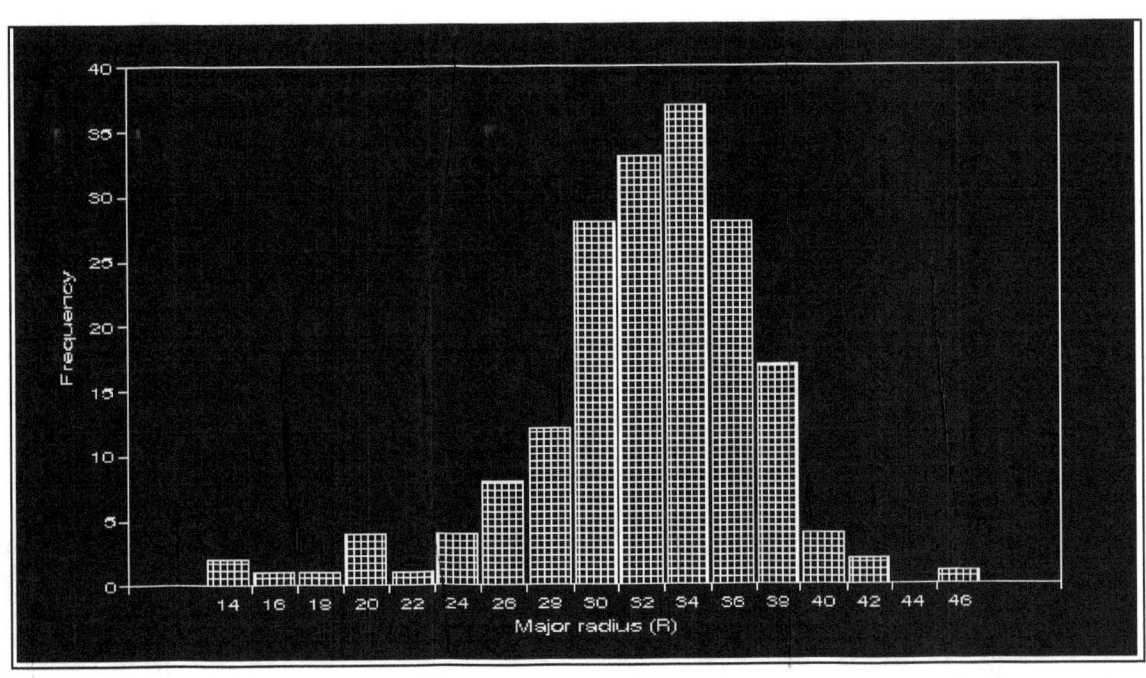

Figure 5.1b *Fromia elegans*
Frequency distribution of major radius / minor radius (R/r)

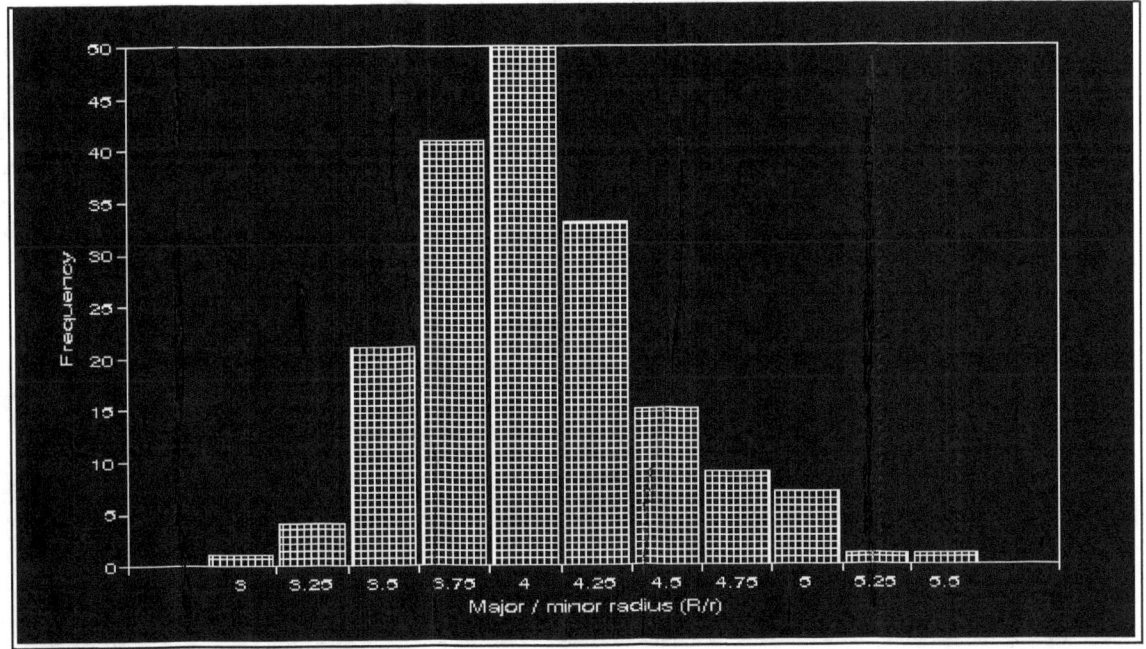

Figure 5.1c *Fromia elegans*
Relation between minor radius (r mm) and major radius (R mm)

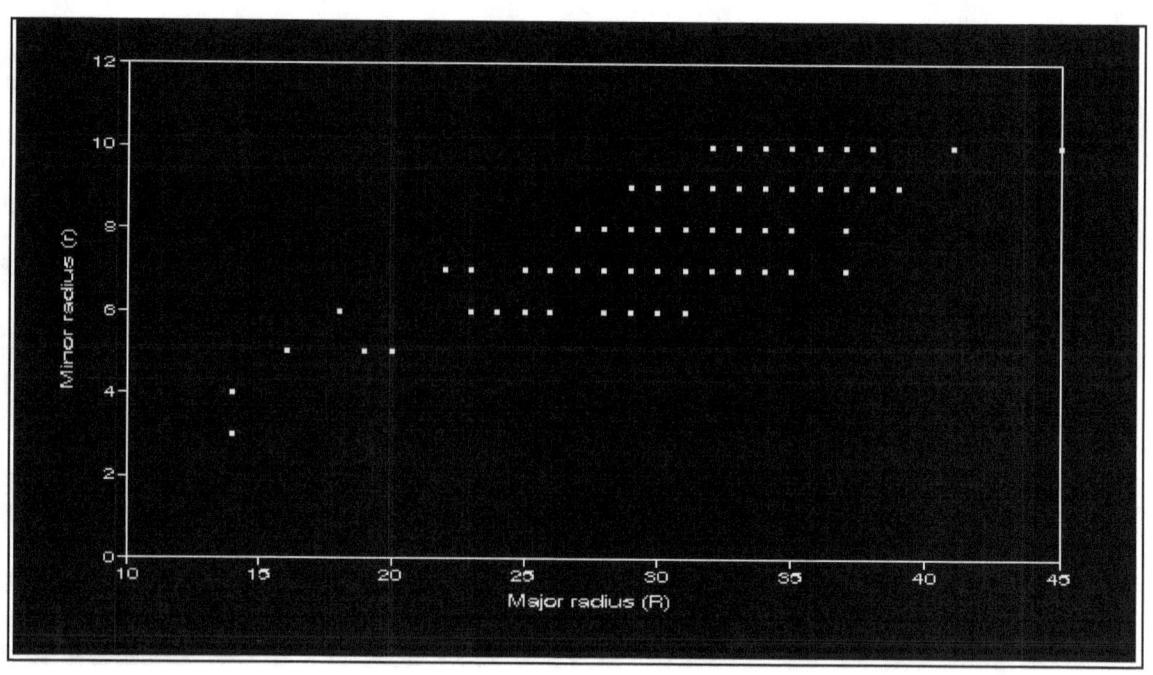

Figure 5.1d *Fromia elegans*
Relation between major radius / minor radius (R/r) and major radius (R mm)

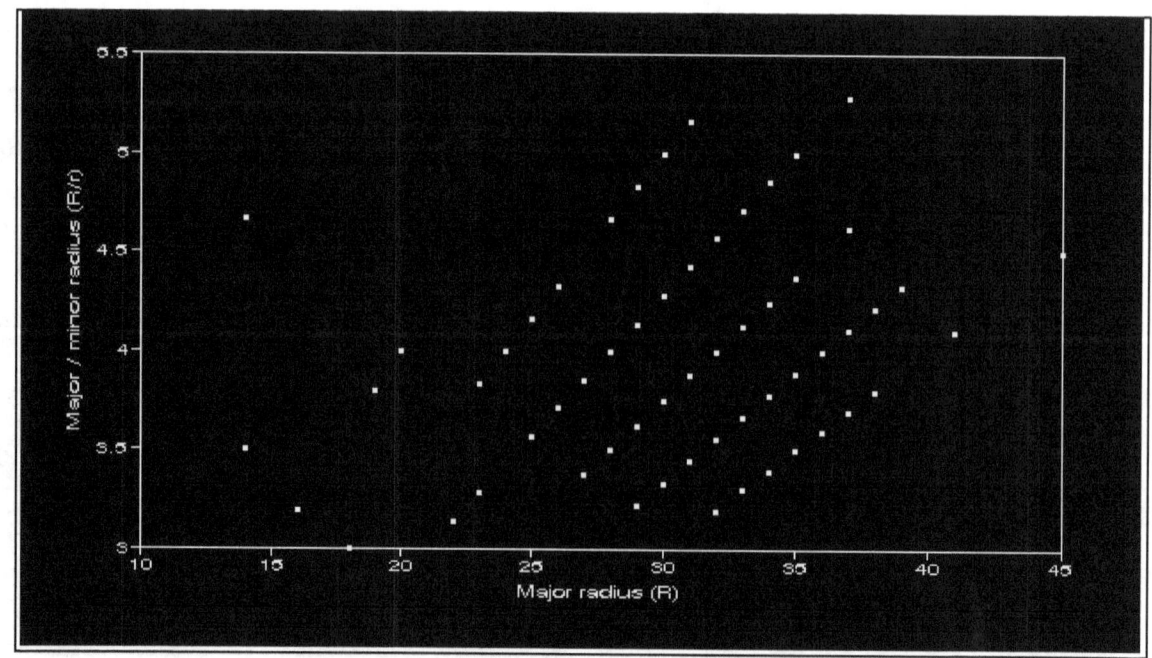

Table 5.2 *Linckia guildingii*

The mean (R mm and R/r), standard deviation (R mm and R/r) and sample size (N) for each sampling period and a grand mean of both R (mm) and R/r, and total sample size are tabled.

PERIOD	MEAN(R)	S.D.(R)	MEAN(R/r)	S.D.(R/r)	N
MAY 1978	117.50	55.45	9.35	1.24	4
AUG 1978	116.25	36.31	-	-	24
NOV 1978	128.00	0.00	-	-	1
JUN 1979	147.00	37.21	10.39	1.22	12
SEP 1979	157.22	31.74	10.21	1.26	9
DEC 1979	133.67	30.39	9.95	1.47	12
APR 1980	134.73	33.74	9.40	1.65	11
JUL 1980	152.60	28.42	9.85	1.97	5
NOV 1980	128.33	8.87	8.61	0.76	6
JUL 1981	139.14	53.28	9.80	1.49	7
JAN 1982	148.53	20.84	10.00	1.17	17
MAY 1982	105.75	37.95	9.37	1.21	4
OCT 1982	114.88	46.06	9.06	0.61	8
DEC 1982	141.64	47.28	10.42	1.12	11
TOTAL	134.23		9.81		131

********** ANOVA **********

Variation in mean Major Radius (R) with respect to time.

Source of Variation	df	SS	MS	F
Among	11.0	20258.8	1841.7	1.4
Within	80.0	105202.5	1315.0	
Total	91.0	125461.3		

P = not significant

Figure 5.2a *Linckia guildingii*
Frequency distribution of major radius (R mm).

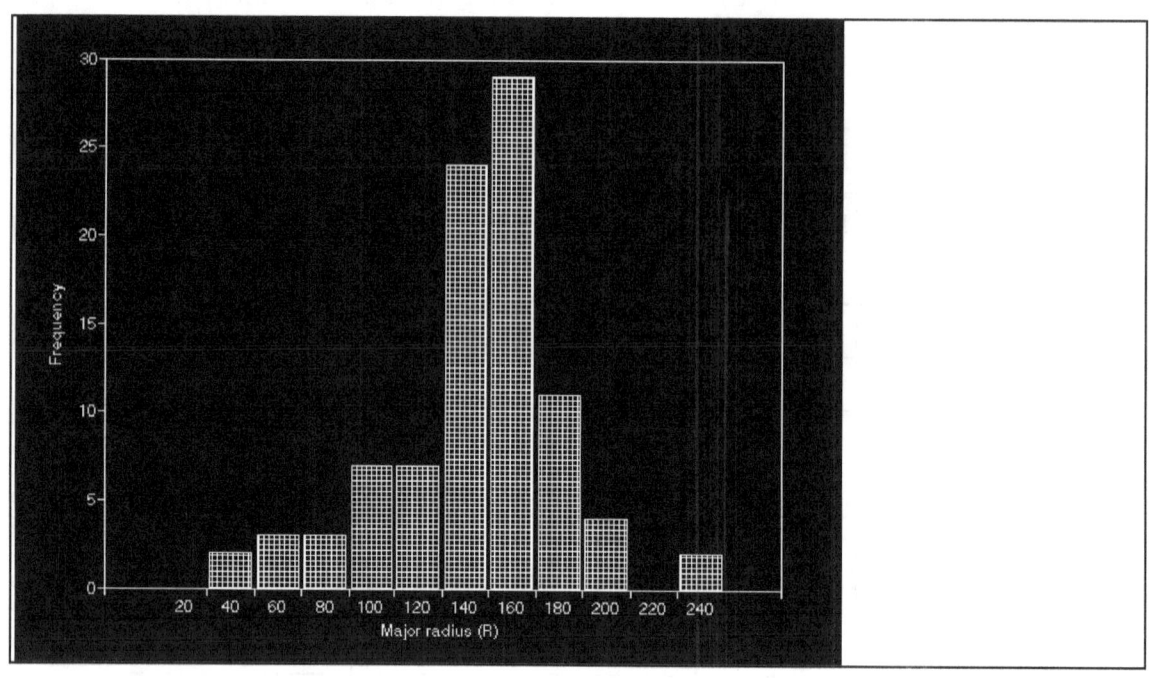

Figure 5.2b *Linckia guildingii*
Frequency distribution of major radius / minor radius (R/r)

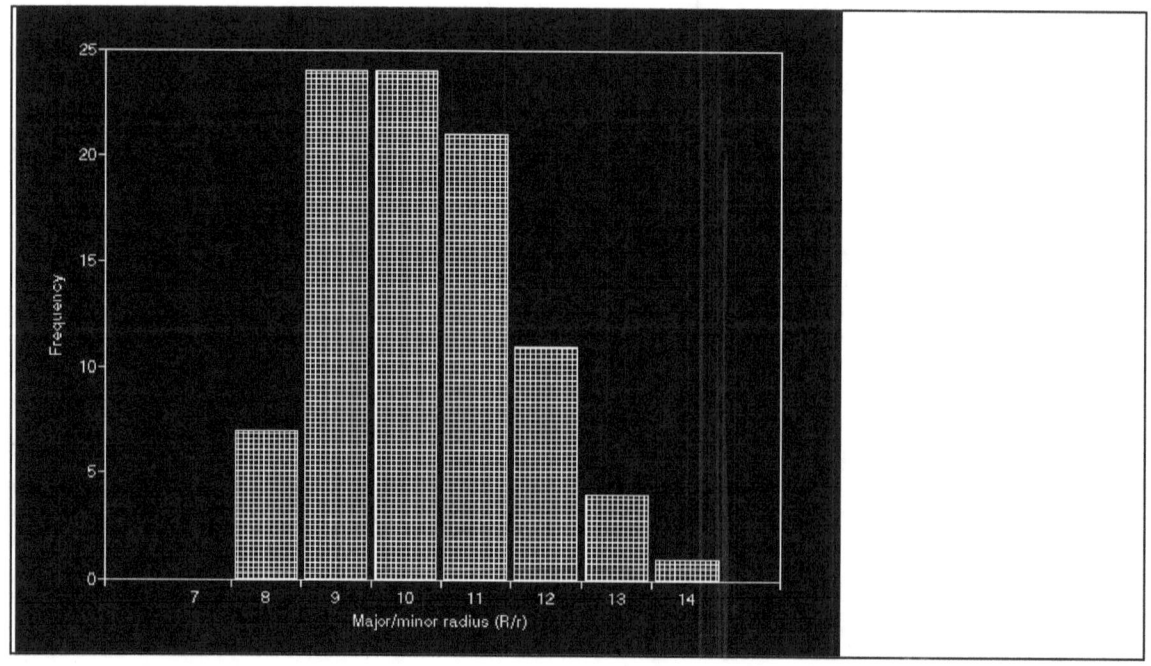

Figure 5.2c *Linckia guildingii*
Relation between minor radius (r mm) and major radius (R mm)

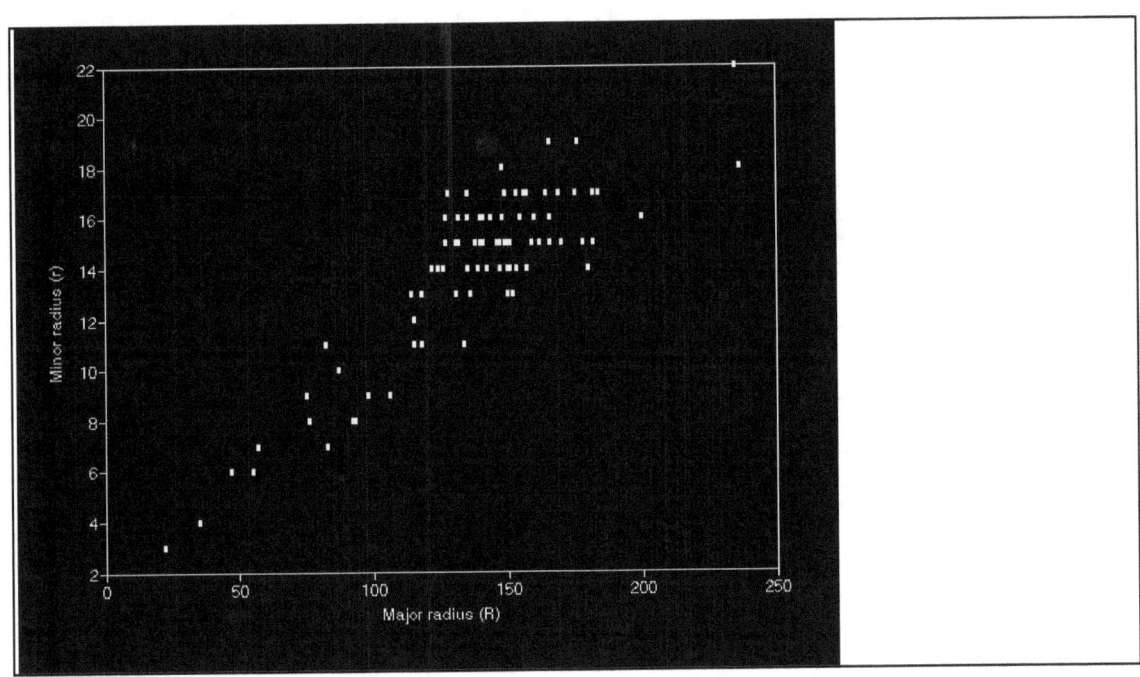

Figure 5.2d *Linckia guildingii*
Relation between major radius / minor radius (R/r)
and major radius (R mm)

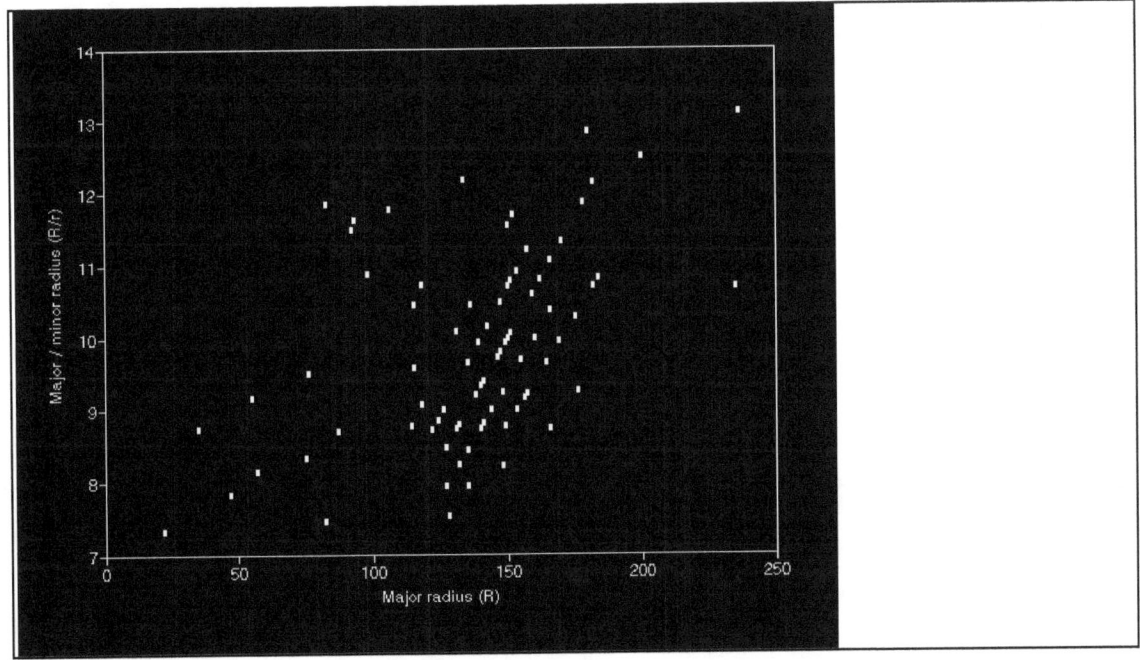

Table 5.3 *Linckia laevigata*
The mean (R mm and R/r), standard deviation (R mm and R/r) and sample size (N) for each sampling period and a grand mean of both R (mm) and R/r, and total sample size are tabled.

PERIOD	MEAN(R)	S.D.(R)	MEAN(R/r)	S.D.(R/r)	N
MAY 1978	126.17	20.45	6.60	0.85	23
AUG 1978	133.28	18.29	-	-	25
FEB 1979	129.55	20.12	-	-	20
JUN 1979	133.13	18.86	6.71	0.73	15
SEP 1979	130.73	19.33	6.64	0.62	11
DEC 1979	134.52	18.13	6.94	0.67	33
APR 1980	131.64	17.92	6.73	0.65	39
JUL 1980	129.78	19.25	6.75	0.64	50
NOV 1980	128.64	21.58	6.55	0.60	44
JUL 1981	126.25	21.81	6.64	0.71	64
JAN 1982	121.77	23.95	6.36	0.63	53
MAY 1982	122.42	18.29	6.29	0.70	65
OCT 1982	121.41	21.74	6.15	0.63	29
DEC 1982	125.07	22.11	6.48	0.60	45
TOTAL	127.17		6.54		516

********** ANOVA **********

Variation in mean Major Radius (R) with respect to time.

Source of Variation	df	SS	MS	F
Among	11.0	7834.0	712.2	1.7
Within	459.0	194746.5	424.3	
Total	470.0	202580.5		

$p < .05$

Figure 5.3a *Linckia laevigata*
Frequency distribution of major radius (R mm).

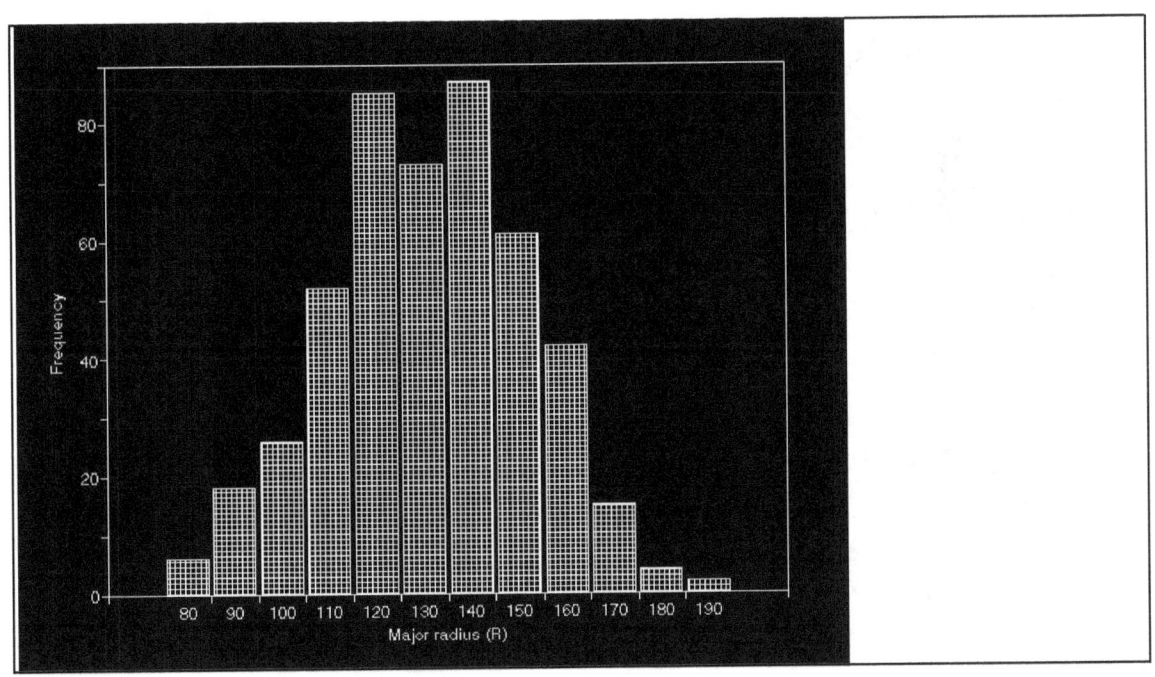

Figure 5.3b *Linckia laevigata*
Frequency distribution of major radius / minor radius (R/r)

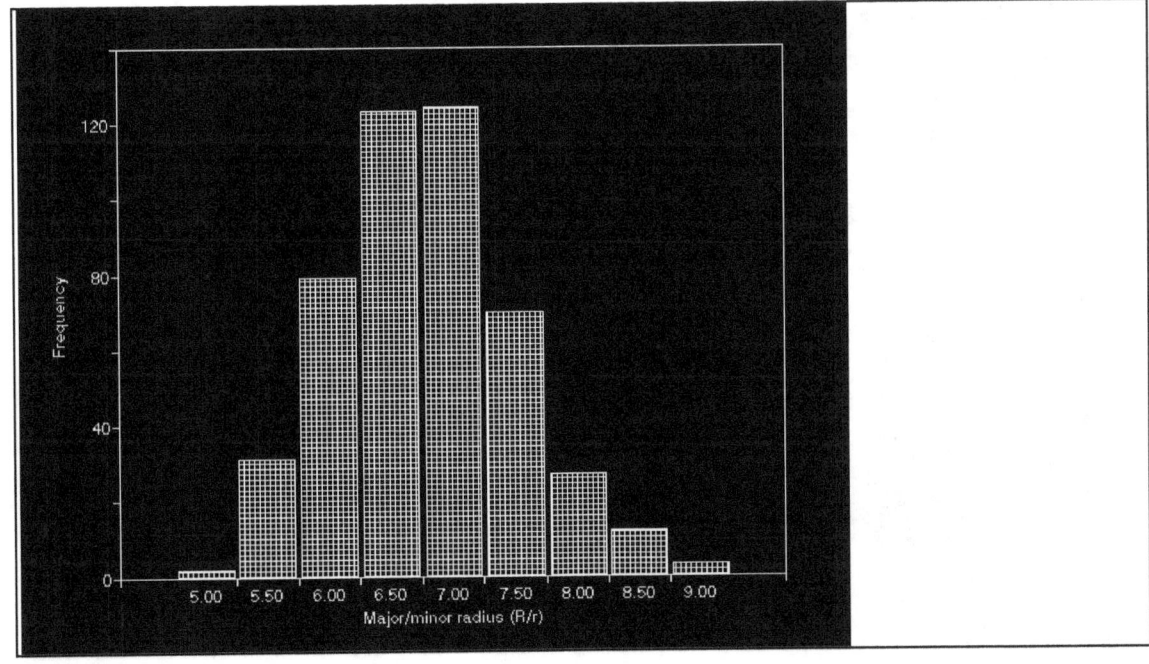

Figure 5.3c *Linckia laevigata*
Relation between minor radius (r mm) and major radius (R mm)

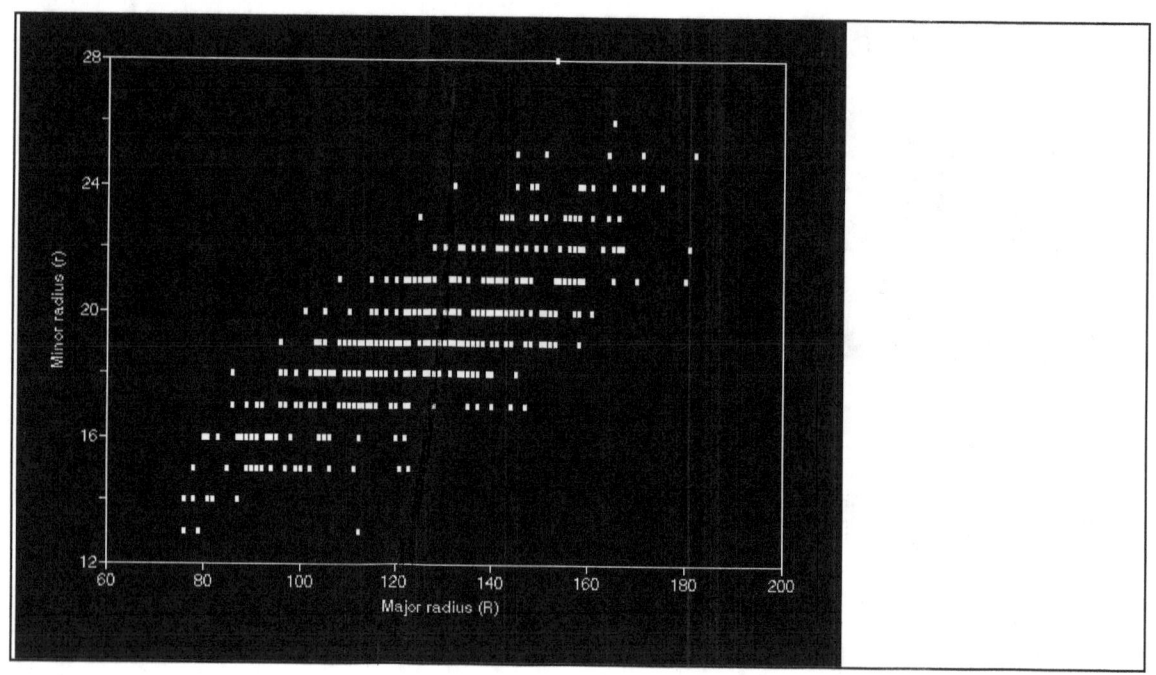

Figure 5.3d *Linckia laevigata*
Relation between major radius / minor radius (R/r)
and major radius (R mm)

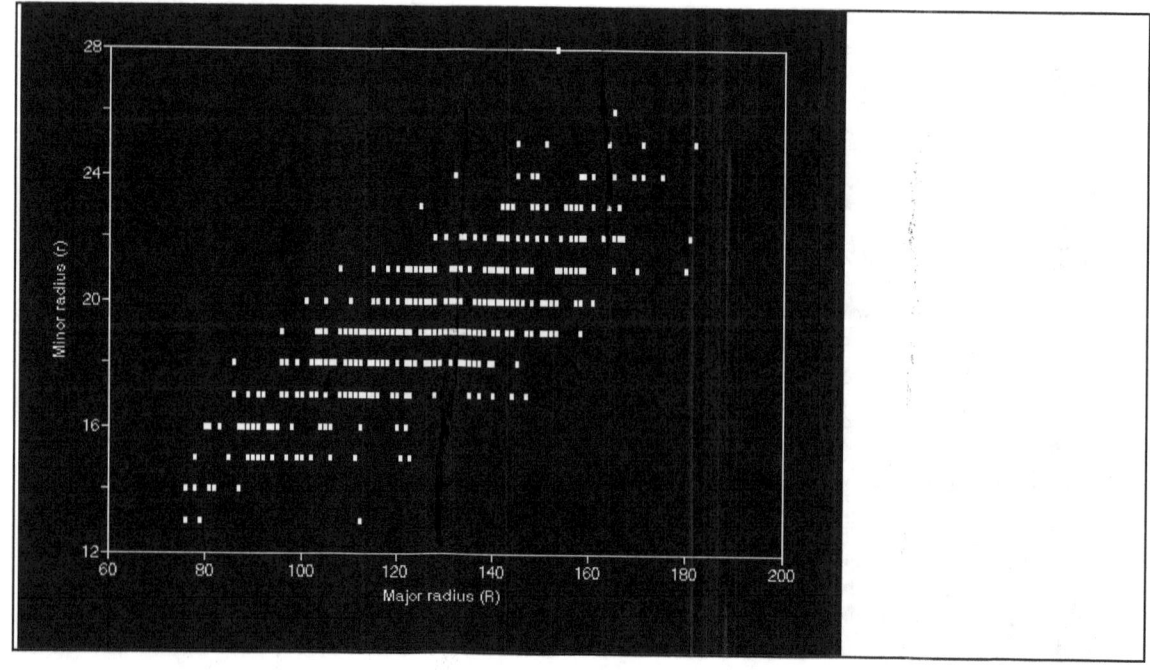

Table 5.4 _Linckia multifora_

The mean (R mm and R/r), standard deviation (R mm and R/r) and sample size (N) for each sampling period and a grand mean of both R (mm) and R/r, and total sample size are tabled.

PERIOD	MEAN(R)	S.D.(R)	MEAN(R/r)	S.D.(R/r)	SAMPLE
MAY 1978	35.45	10.10	7.00	1.37	56
AUG 1978	37.20	7.80	-	-	44
NOV 1978	38.60	11.60	-	-	12
JUN 1979	48.46	12.76	7.75	1.23	54
SEP 1979	41.68	10.61	7.63	0.91	19
DEC 1979	32.67	10.69	6.48	1.17	3
JUL 1980	45.41	17.15	7.15	1.46	17
NOV 1980	39.74	10.51	7.16	1.15	23
JUL 1981	31.34	7.74	7.78	1.56	50
JAN 1982	35.17	8.93	7.21	1.36	30
MAY 1982	34.20	14.73	6.70	1.46	15
OCT 1982	35.90	5.36	7.08	1.36	21
DEC 1982	36.27	8.77	6.89	0.99	52
TOTAL	38.01		7.27		396

*********** ANOVA **********

Variation in mean Major Radius (R) with respect to time.

Source of Variation	df	SS	MS	F
Among	10.0	10543.4	1054.3	9.5
Within	329.0	36414.4	110.7	
Total	339.0	46957.8		

$P < .001$

Figure 5.4a *Linckia multifora*
Frequency distribution of major radius (R mm).

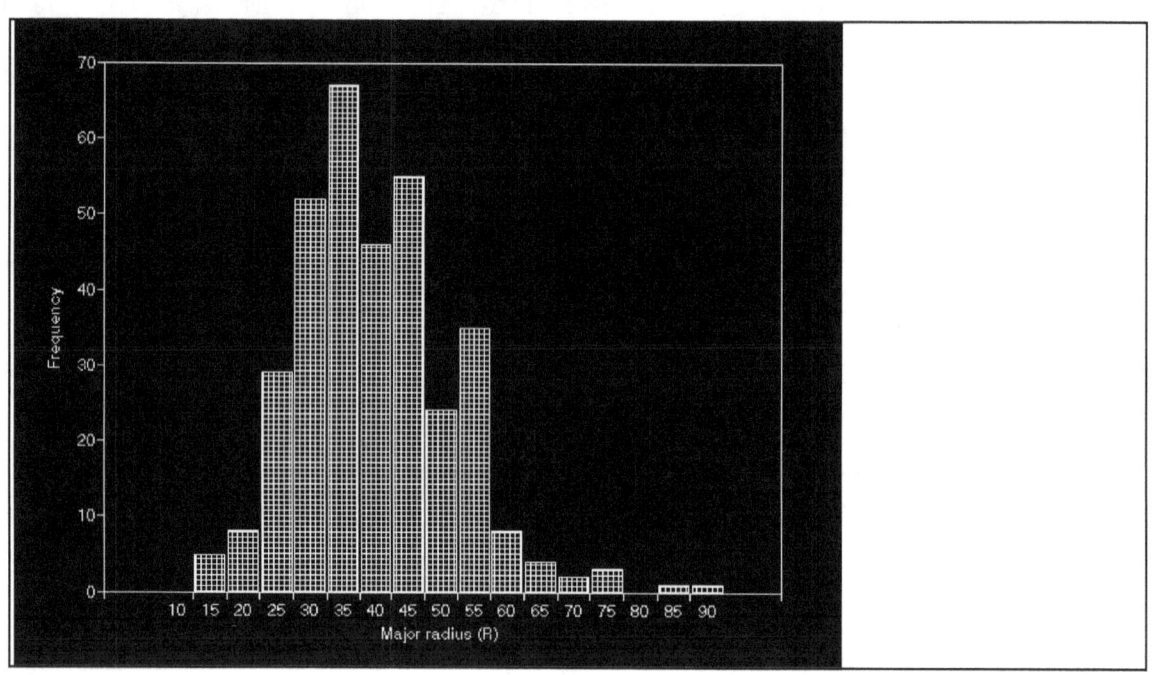

Figure 5.4b *Linckia multifora*
Frequency distribution of major radius / minor radius (R/r)

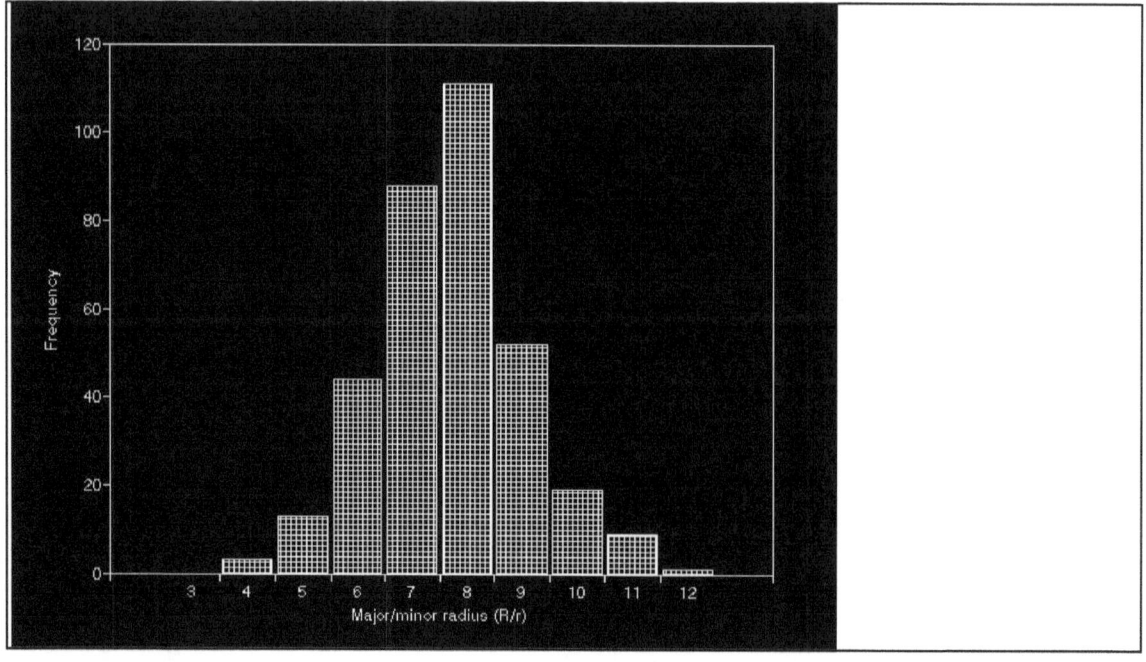

Figure 5.4c *Linckia multifora*
Relation between minor radius (r mm) and major radius (R mm)

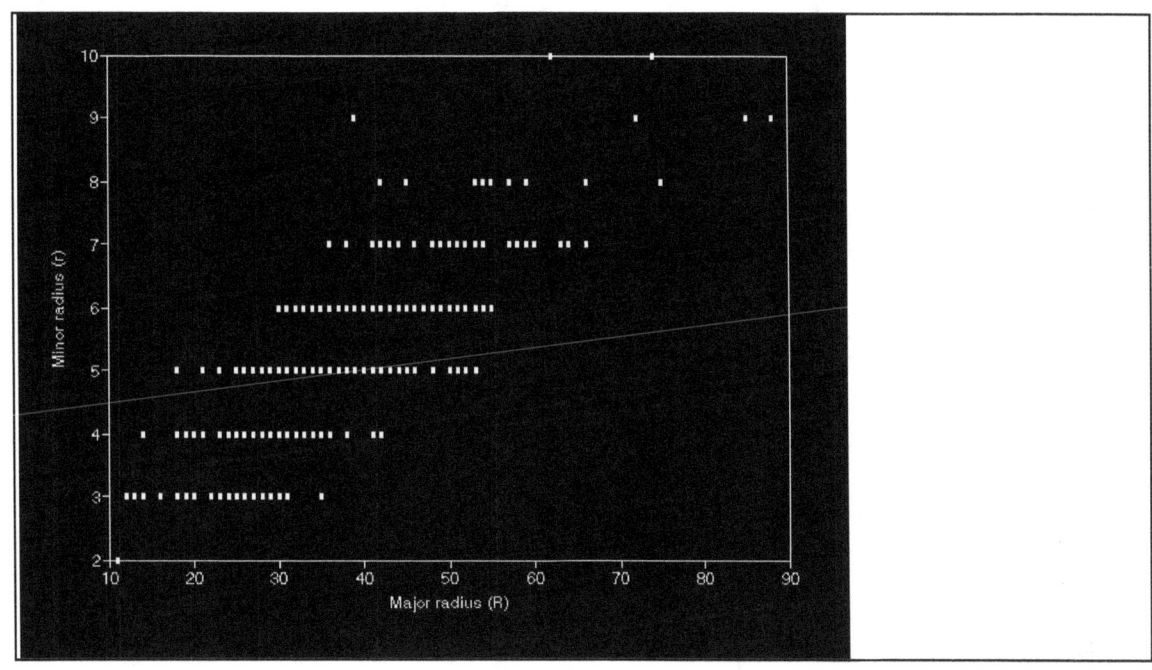

Figure 5.4d *Linckia multifora*
Relation between major radius / minor radius (R/r)
and major radius (R mm)

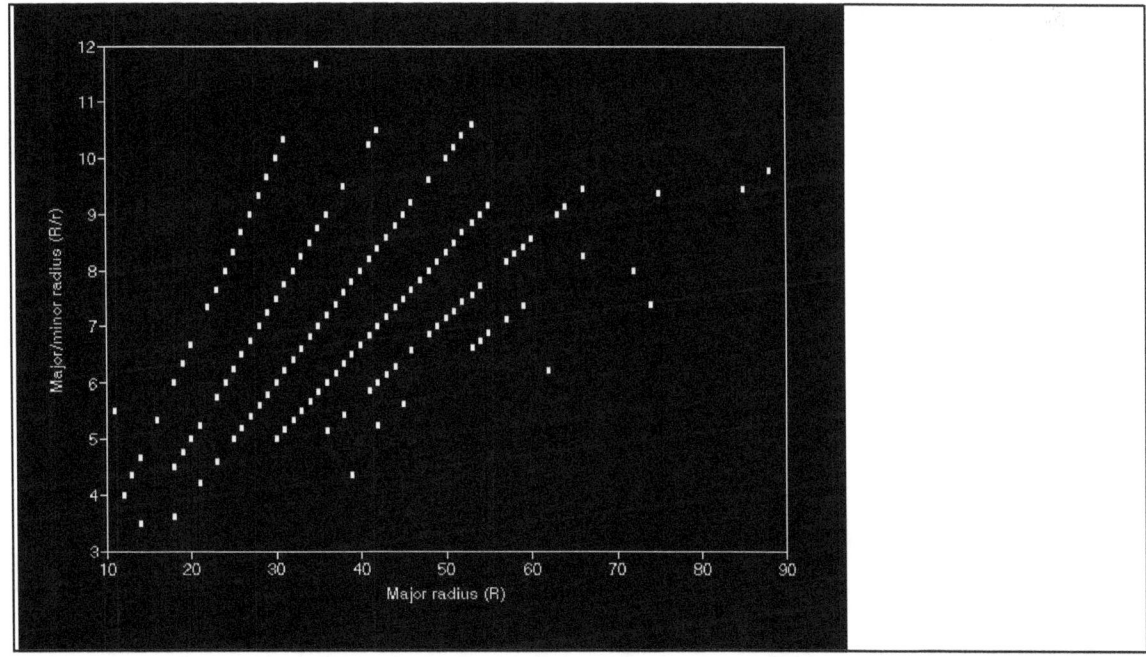

Table 5.5 *Nardoa novaecaledoniae*

The mean (R mm and R/r), standard deviation (R mm and R/r) and sample size (N) for each sampling period and a grand mean of both R (mm) and R/r, and total sample size are tabled.

PERIOD	MEAN(R)	S.D.(R)	MEAN(R/r)	S.D.(R/r)	SAMPLE
MAY 1978	83.00	11.07	5.20	0.50	17
AUG 1978	86.03	15.05	-	-	32
FEB 1979	93.27	10.22	-	-	22
JUN 1979	86.38	13.54	5.58	0.76	16
SEP 1979	86.42	13.57	5.70	0.55	19
DEC 1979	88.80	10.39	5.68	0.54	25
APR 1980	88.94	11.52	5.52	0.66	33
JUL 1980	82.79	12.79	5.60	0.53	33
NOV 1980	81.74	9.21	5.34	0.59	35
JUL 1981	87.50	9.07	5.94	0.69	37
JAN 1982	90.05	12.44	5.85	0.50	20
MAY 1982	92.95	14.25	5.91	0.48	21
OCT 1982	92.21	12.25	5.85	0.55	29
DEC 1982	94.09	10.85	5.67	0.42	22
TOTAL	87.87		5.65		361

*********** ANOVA **********

Variation in mean Major Radius (R) with respect to time.

```
Source of Variation        df        SS        MS        F
Among                     11.0    4732.8     430.3      3.2
Within                   295.0   39614.3     134.3
Total                    306.0   44347.0
P < .01
```

Figure 5.5a *Nardoa novaecaledoniae*
Frequency distribution of major radius (R mm).

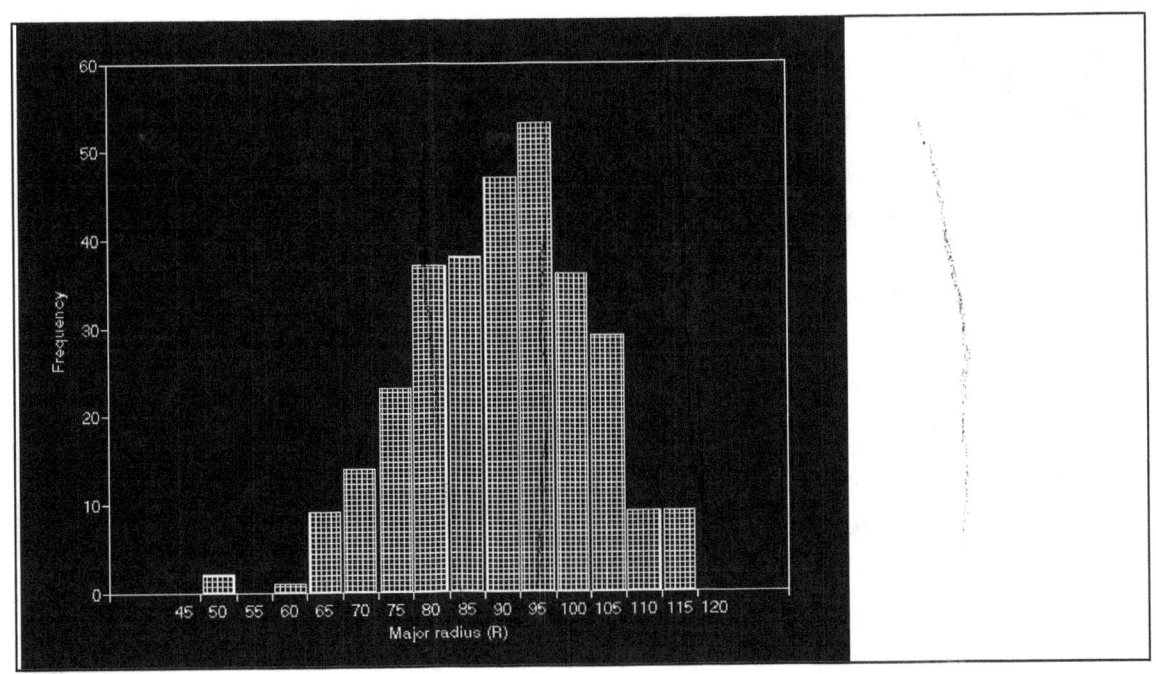

Figure 5.5b *Nardoa novaecaledoniae*
Frequency distribution of major radius / minor radius (R/r)

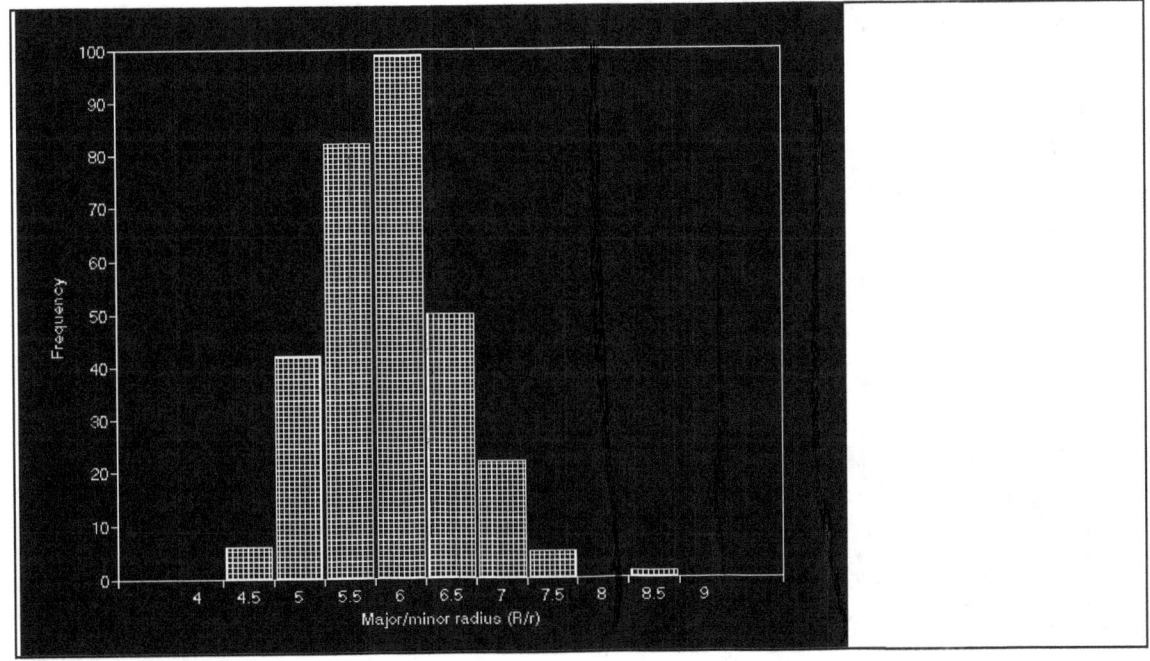

Figure 5.5c *Nardoa novaecaledoniae*
Relation between minor radius (r mm) and major radius (R mm)

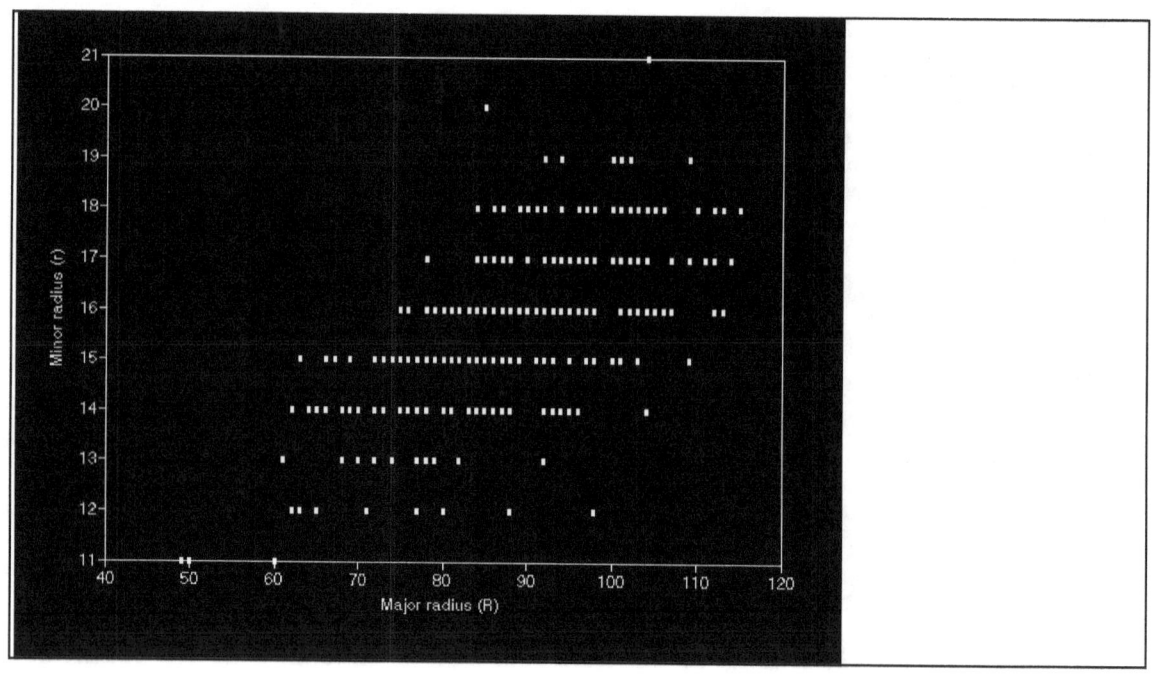

Figure 5.5d *Nardoa novaecaledoniae*
Relation between major radius / minor radius (R/r)
and major radius (R mm)

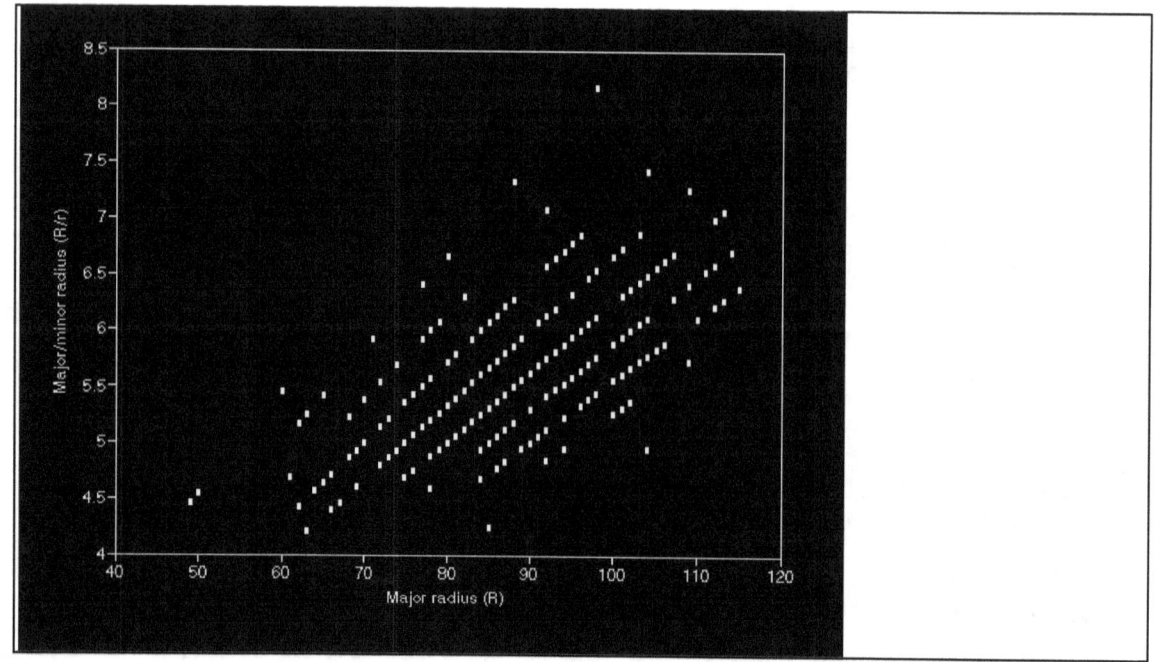

Table 5.6 *Nardoa pauciforis*

The mean (R mm and R/r), standard deviation (R mm and R/r) and sample size (N) for each sampling period and a grand mean of both R (mm) and R/r, and total sample size are tabled.

PERIOD	MEAN(R)	S.D.(R)	MEAN(R/r)	S.D.(R/r)	N
MAY 1978	97.89	9.53	5.90	0.69	9
AUG 1978	104.00	10.86	-	-	16
FEB 1979	104.92	10.77	-	-	26
JUN 1979	97.82	8.00	6.27	0.51	11
SEP 1979	107.89	10.94	6.52	0.66	9
DEC 1979	99.50	9.90	6.54	0.57	12
APR 1980	103.86	11.77	6.52	0.60	21
JUL 1980	100.74	13.14	6.60	0.66	19
NOV 1980	100.00	9.92	6.66	0.65	16
JUL 1981	104.83	12.60	6.79	0.69	30
JAN 1982	99.94	10.24	6.48	0.78	18
MAY 1982	109.09	7.49	6.75	0.44	11
OCT 1982	108.27	4.40	6.82	0.76	15
DEC 1982	108.90	13.67	6.69	0.61	20
TOTAL	103.66		6.59		233

********** ANOVA **********

Variation in mean Major Radius (R) with respect to time.

Source of Variation	df	SS	MS	F
Among	11.0	2896.4	263.3	2.2
Within	179.0	21529.3	120.3	
Total	190.0	24425.6		

$p < .05$

Figure 5.6a *Nardoa pauciforis*
Frequency distribution of major radius (R mm).

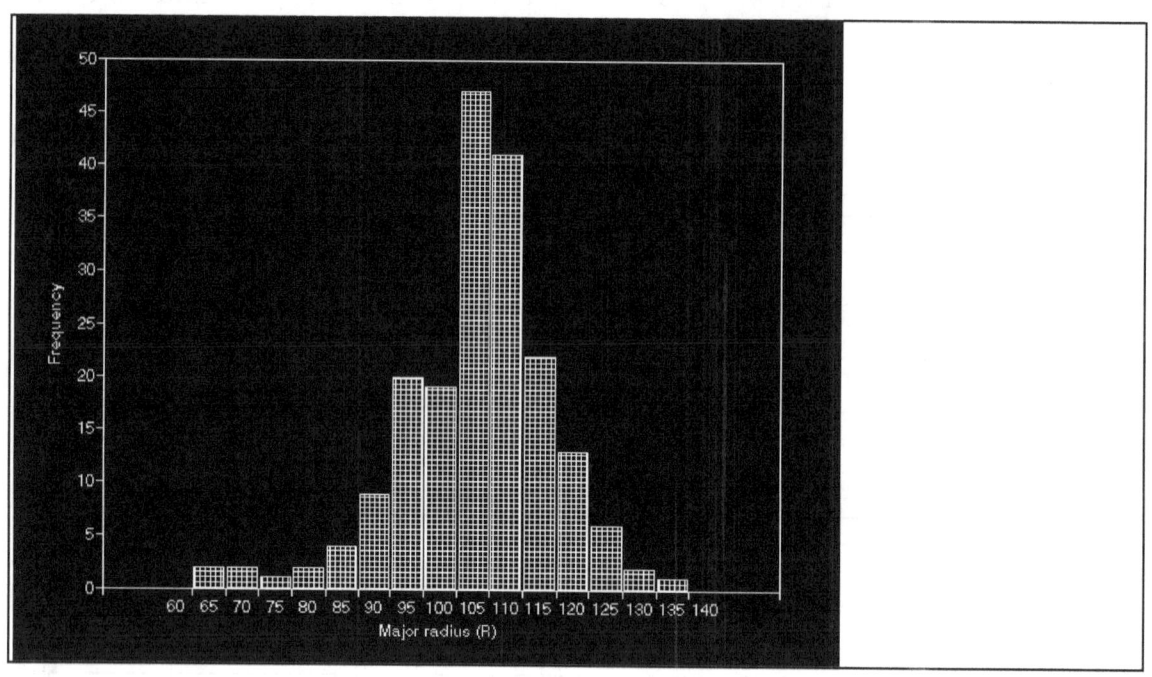

Figure 5.6b *Nardoa pauciforis*
Frequency distribution of major radius / minor radius (R/r)

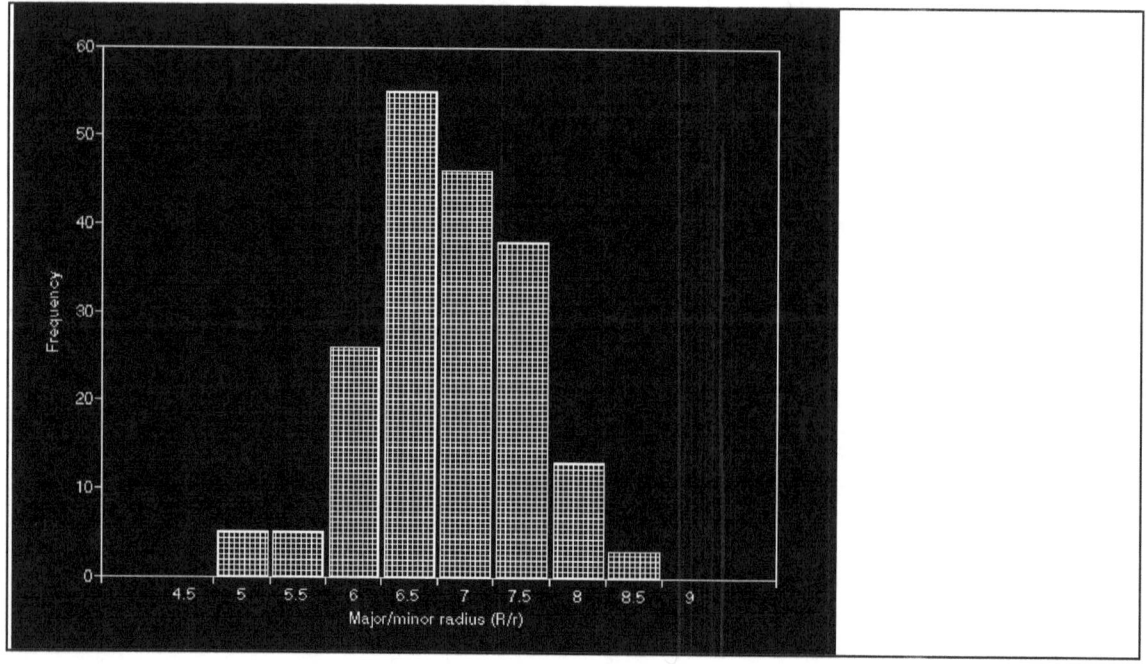

Figure 5.6c *Nardoa pauciforis*
Relation between minor radius (r mm) and major radius (R mm)

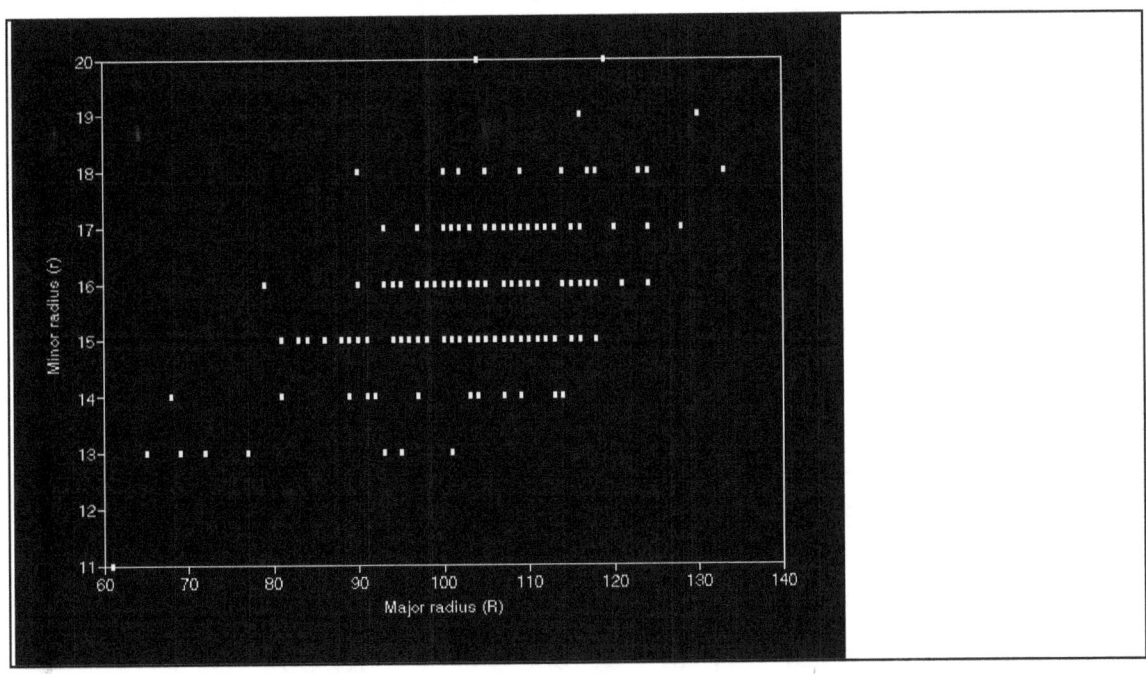

Figure 5.6d *Nardoa pauciforis*
Relation between major radius / minor radius (R/r)
and major radius (R mm)

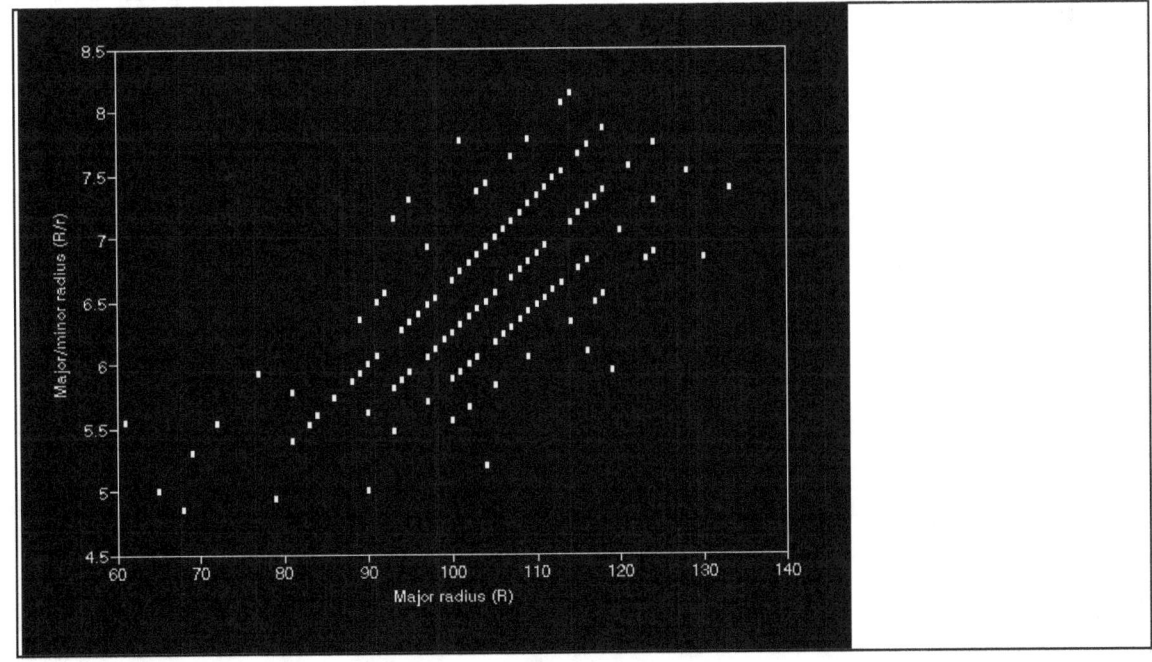

Table 5.7 *Ophidiaster granifer*

The mean (R mm and R/r), standard deviation (R mm and R/r) and sample size (N) for each sampling period and a grand mean of both R (mm) and R/r, and total sample size are tabled.

PERIOD	MEAN(R)	S.D.(R)	MEAN(R/r)	S.D.(R/r)	N
MAY 1978	25.94	3.88	4.41	0.45	17
AUG 1978	25.50	5.63	-	-	12
NOV 1978	22.10	6.89	-	-	10
JUN 1979	29.83	4.22	4.37	0.47	6
SEP 1979	25.00	0.00	3.57	0.00	1
DEC 1979	28.56	4.82	4.67	0.45	9
APR 1980	33.00	2.16	4.27	0.40	4
JUL 1980	31.80	3.70	4.54	0.53	5
NOV 1980	28.25	2.63	4.35	0.25	4
JUL 1981	22.17	2.71	4.85	0.92	6
JAN 1982	26.90	3.28	4.48	0.24	10
MAY 1982	29.50	9.65	4.74	0.49	8
OCT 1982	25.88	5.63	4.08	0.34	16
DEC 1982	27.69	5.68	3.74	0.48	13
TOTAL	26.81		4.34		121

********** ANOVA **********

Variation in mean Major Radius (R) with respect to time.

Source of Variation	df	SS	MS	F
Among	10.0	548.7	54.9	2.1
Within	87.0	2253.8	25.9	
Total	97.0	2802.5		

P = not significant

Figure 5.7 *Ophidiaster granifer*
Frequency distribution of major radius (R mm).

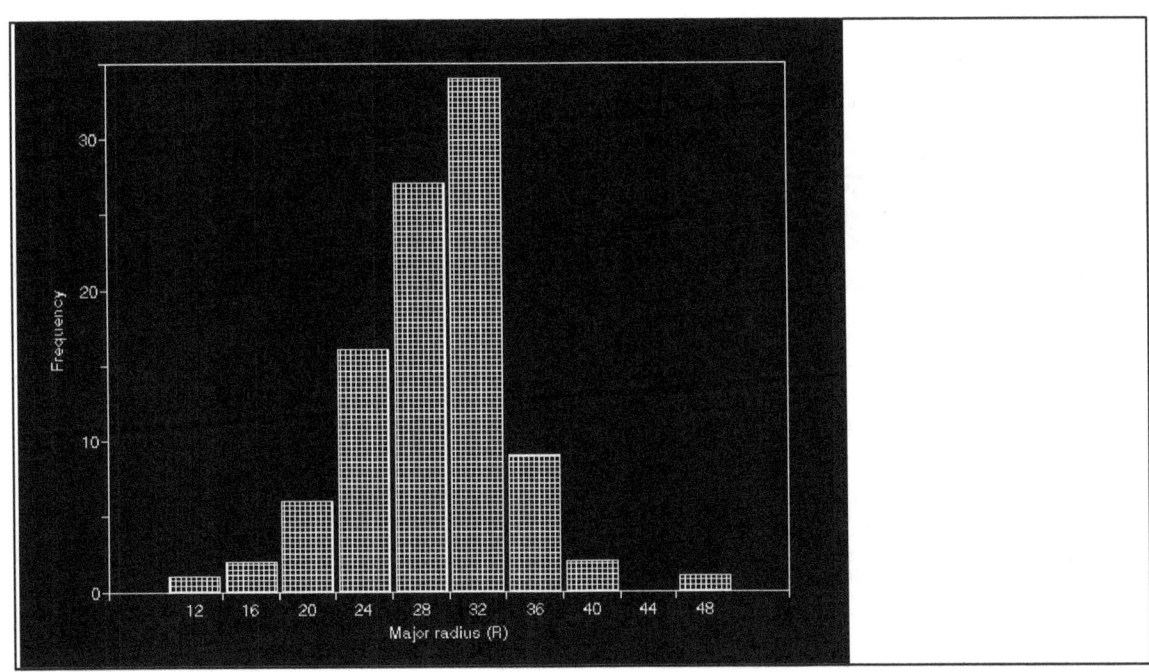

Figure 5.7 *Ophidiaster granifer*
Frequency distribution of major radius / minor radius (R/r)

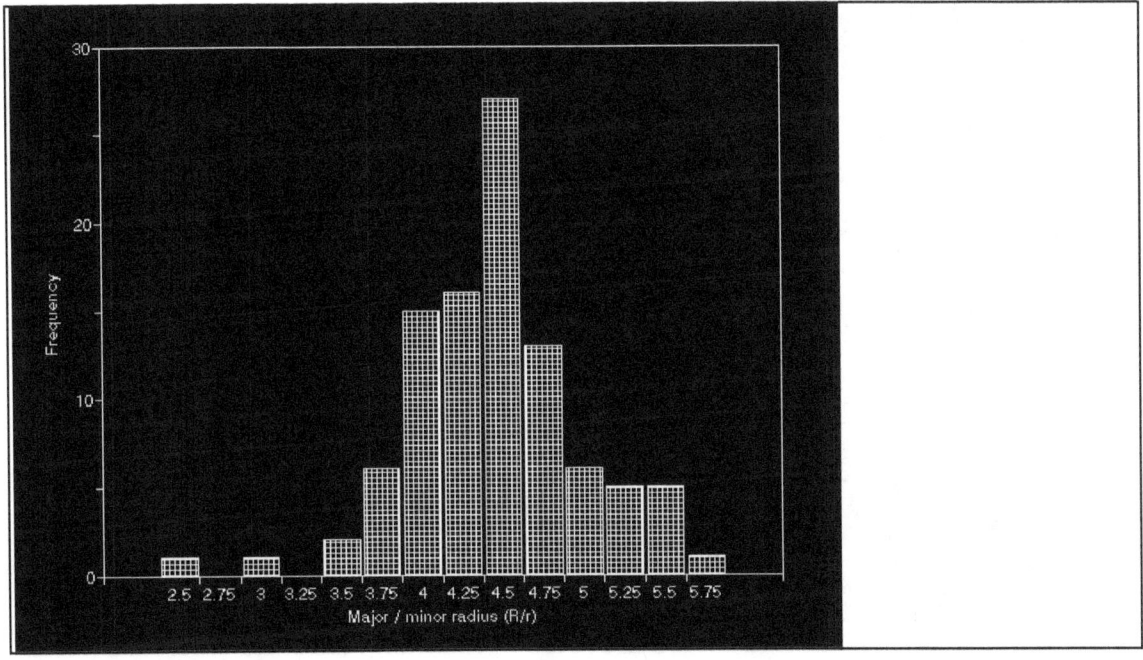

Figure 5.7 *Ophidiaster granifer*
Relation between minor radius (r mm) and major radius (R mm)

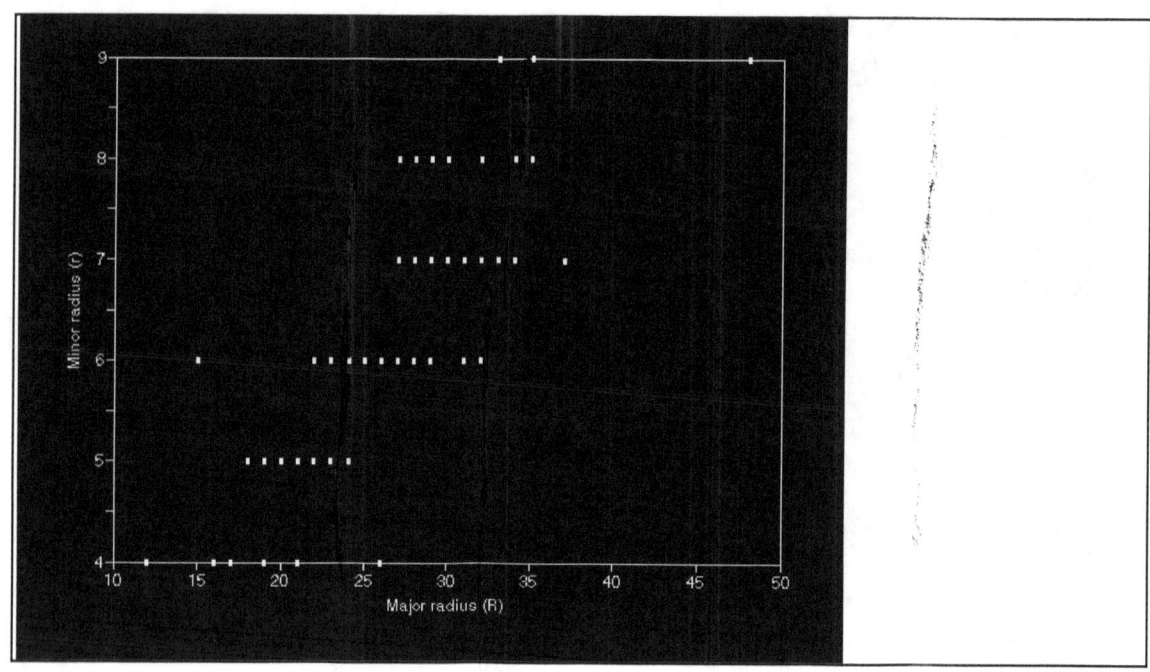

Figure 5.7 *Ophidiaster granifer*
Relation between major radius / minor radius (R/r)
and major radius (R mm)

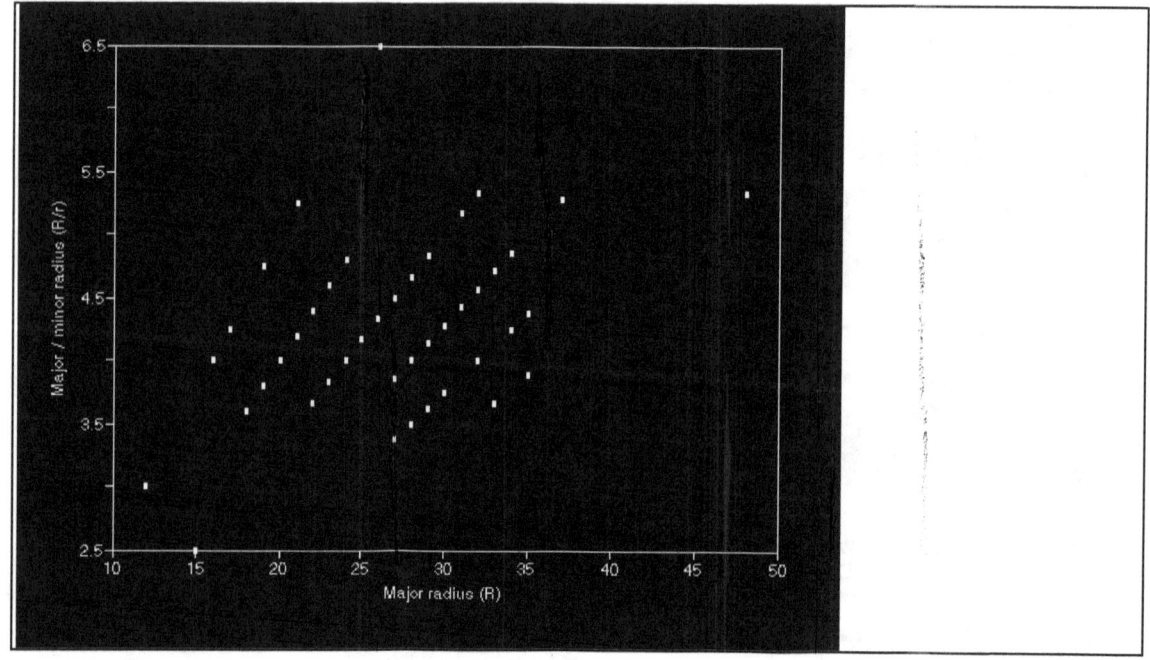

Table 5.8 *Asterina burtoni*

The mean (R mm and R/r), standard deviation (R mm and R/r) and sample size (N) for each sampling period and a grand mean of both R (mm) and R/r, and total sample size are tabled.

PERIOD	MEAN(R)	S.D.(R)	MEAN(R/r)	S.D.(R/r)	N
MAY 1978	11.91	3.75	1.83	0.25	42
AUG 1978	11.66	3.30	-	-	58
NOV 1978	13.89	5.69	-	-	9
JUN 1979	13.00	4.07	1.89	0.19	11
SEP 1979	12.67	2.52	1.80	0.11	3
APR 1980	14.67	2.66	1.92	0.16	6
JUL 1980	14.69	3.54	1.81	0.24	13
NOV 1980	13.50	5.04	1.85	0.14	8
JUL 1981	11.50	7.78	1.70	0.28	2
JAN 1982	15.50	3.53	1.88	0.16	12
MAY 1982	13.00	6.08	1.90	0.48	3
OCT 1982	12.75	3.70	1.70	0.17	16
DEC 1982	15.10	3.82	1.90	0.23	20
TOTAL	12.92		1.83		203

*********** ANOVA **********

Variation in mean Major Radius (R) with respect to time.

Source of Variation	df	SS	MS	F
Among	10.0	255.8	25.6	1.7
Within	126.0	1886.7	15.0	
Total	136.0	2142.5		

P = not significant

Figure 5.8a *Asterina burtoni*
Frequency distribution of major radius (R mm).

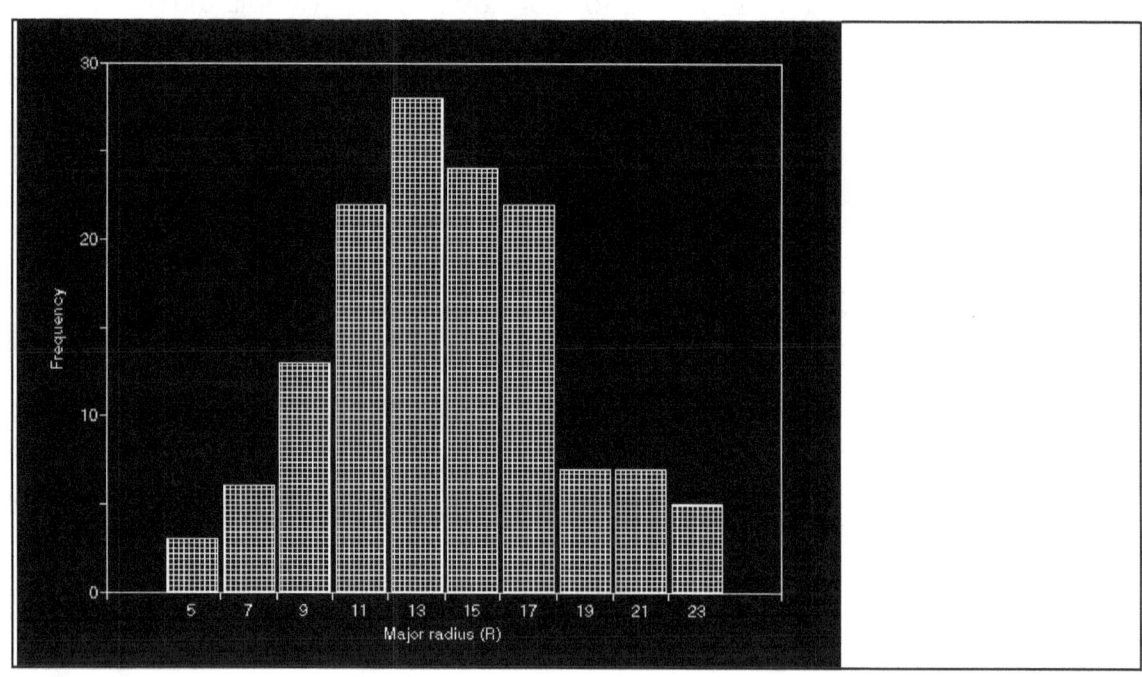

Figure 5.8b *Asterina burtoni*
Frequency distribution of major radius / minor radius (R/r)

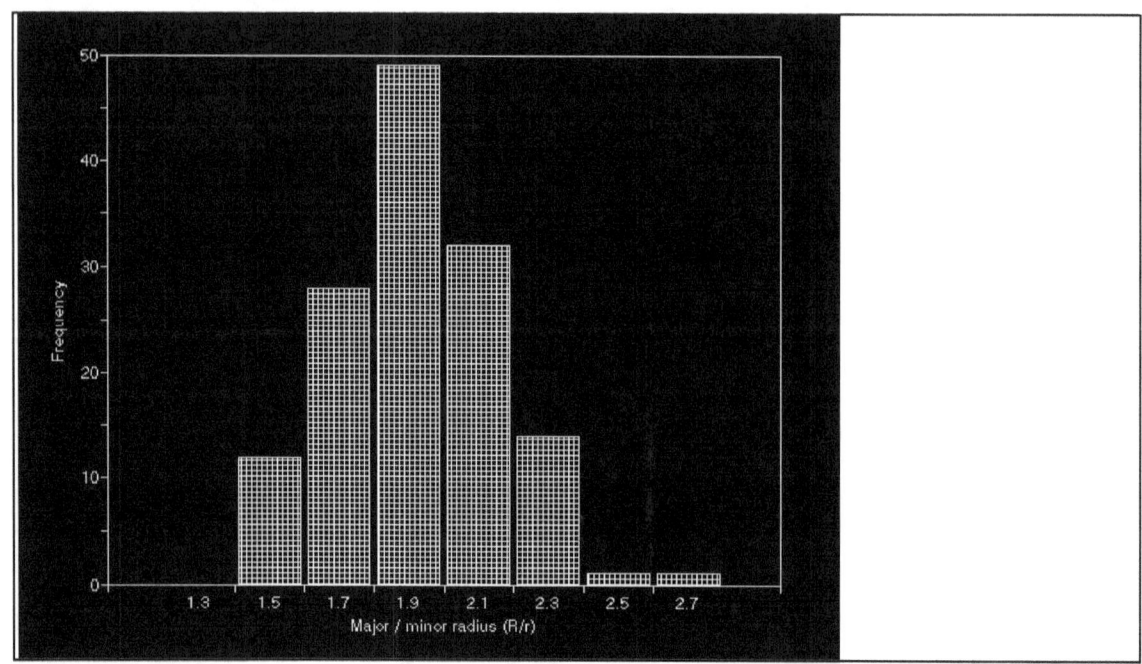

Figure 5.8c *Asterina burtoni*
Relation between minor radius (r mm) and major radius (R mm)

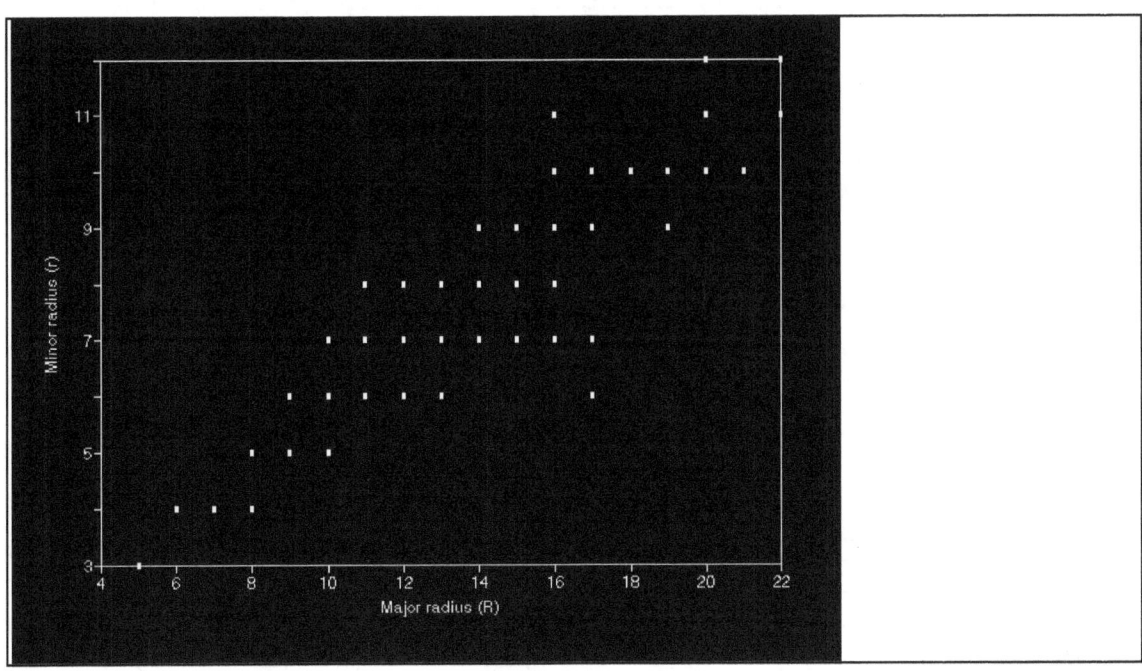

Figure 5.8d *Asterina burtoni*
Relation between major radius / minor radius (R/r)
and major radius (R mm)

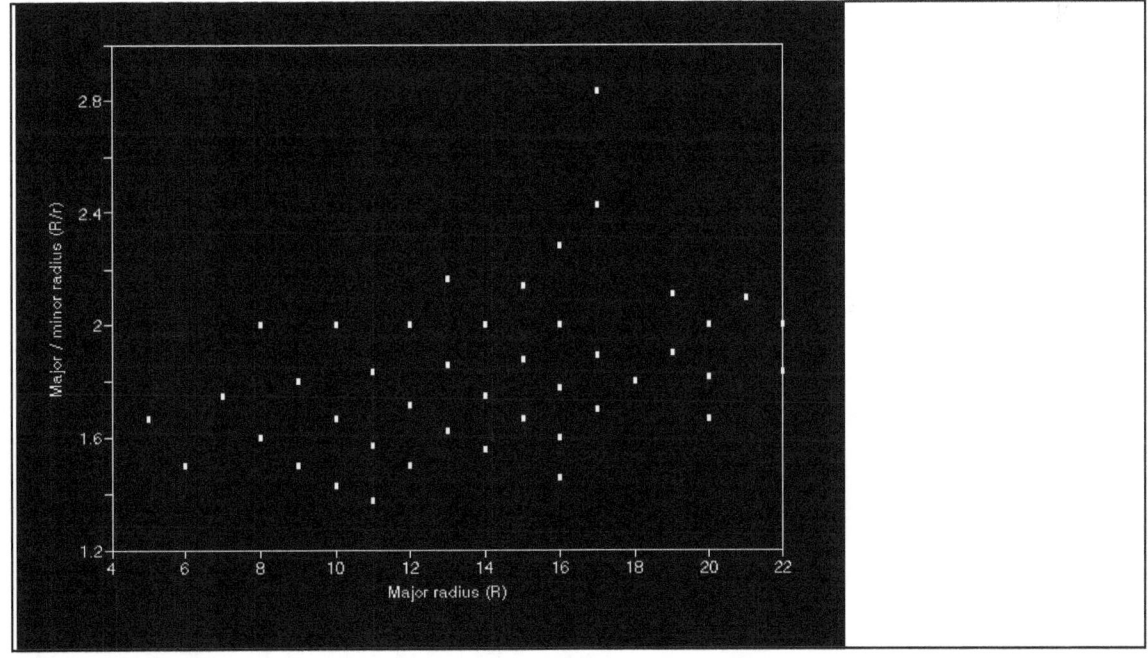

Table 5.9 _Disasterina abnormalis_

The mean (R mm and R/r), standard deviation (R mm and R/r) and sample size (N) for each sampling period and a grand mean of both R (mm) and R/r, and total sample size are tabled.

PERIOD	MEAN(R)	S.D.(R)	MEAN(R/r)	S.D.(R/r)	N
MAY 1978	15.27	4.14	1.82	0.18	92
AUG 1978	15.14	3.76	-	-	57
NOV 1978	15.25	3.51	-	-	20
JUN 1979	15.31	3.42	1.91	0.18	80
SEP 1979	16.41	2.45	1.87	0.15	85
DEC 1979	15.84	2.95	1.90	0.21	215
APR 1980	13.39	3.57	2.06	0.24	161
JUL 1980	13.55	3.56	1.76	0.19	98
NOV 1980	14.64	3.33	1.88	0.31	66
JUL 1981	13.10	2.97	1.94	0.26	20
JAN 1982	13.58	3.75	1.72	0.20	78
MAY 1982	15.76	5.04	2.00	0.25	25
OCT 1982	10.70	4.91	1.75	0.27	79
DEC 1982	16.09	3.26	1.85	0.19	33
TOTAL	14.55		1.87		1109

********** ANOVA *********

Variation in mean Major Radius (R) with respect to time.

Source of Variation	df	SS	MS	F
Among	12.0	2462.0	205.2	15.8
Within	1054.0	13716.7	13.0	
Total	1066.0	16178.7		

$P < .001$

Figure 5.9a *Disasterina abnormalis*
Frequency distribution of major radius (R mm).

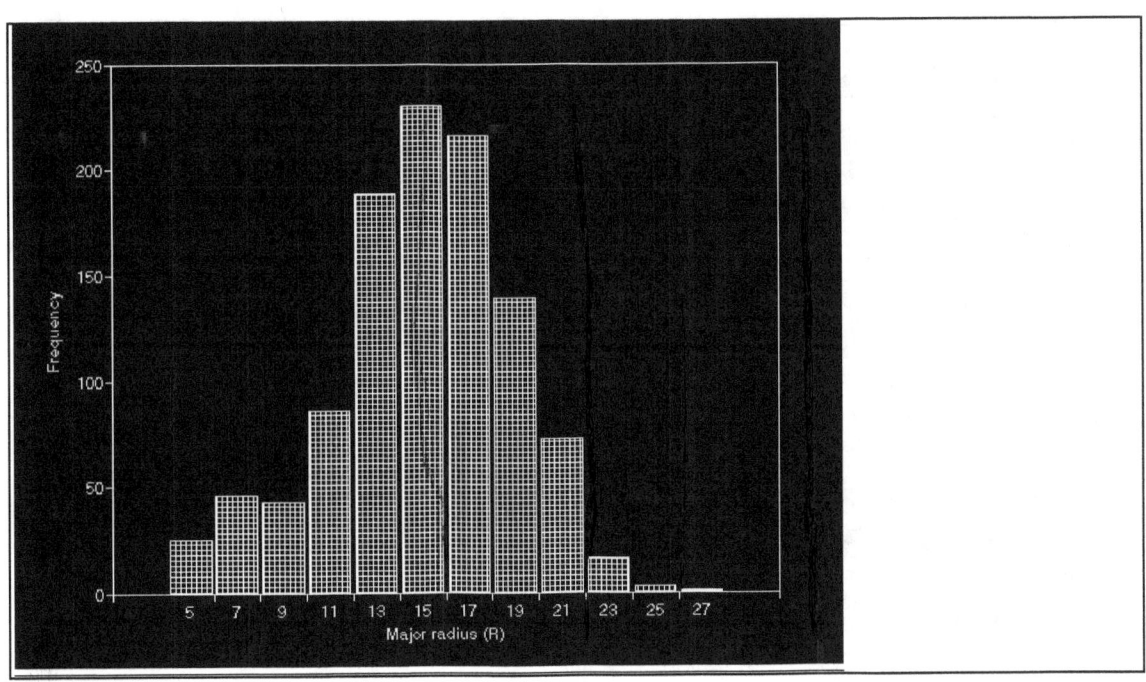

Figure 5.9b *Disasterina abnormalis*
Frequency distribution of major radius / minor radius (R/r)

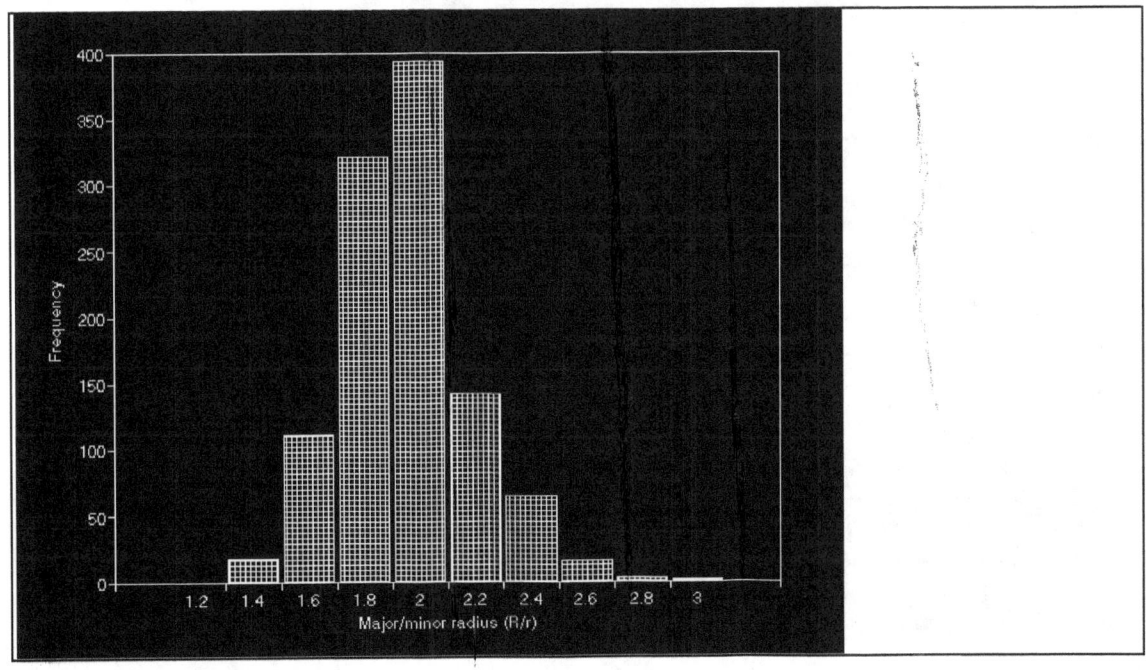

Figure 5.9c *Disasterina abnormalis*
Relation between minor radius (r mm) and major radius (R mm)

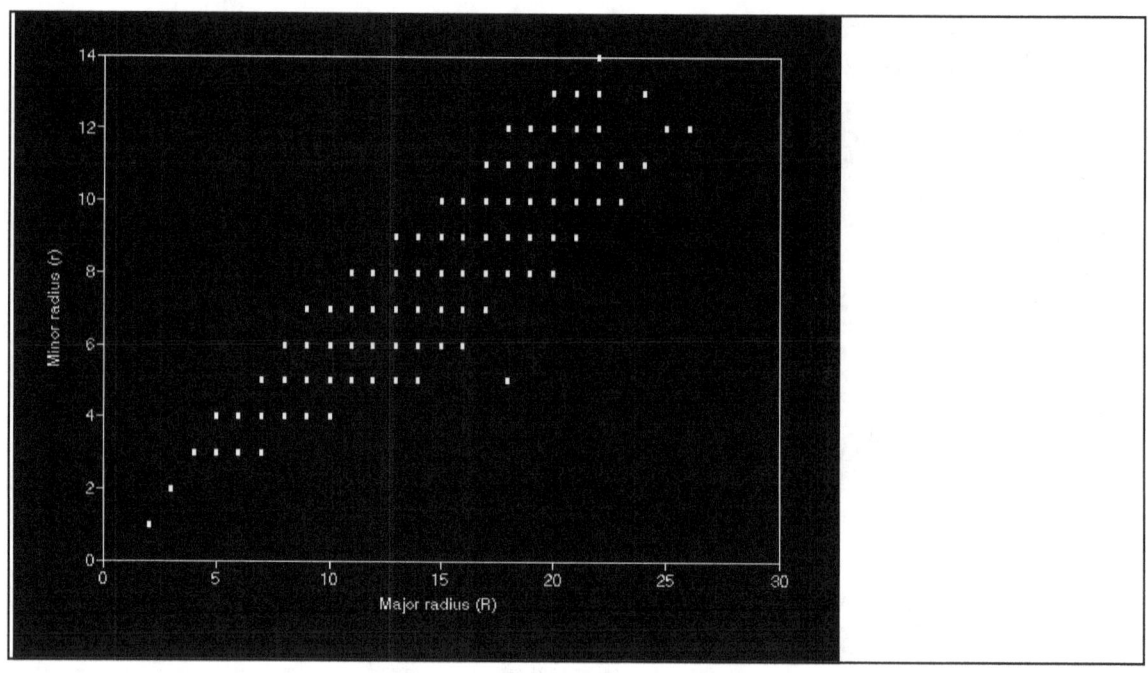

Figure 5.9d *Disasterina abnormalis*
Relation between major radius / minor radius (R/r)
and major radius (R mm)

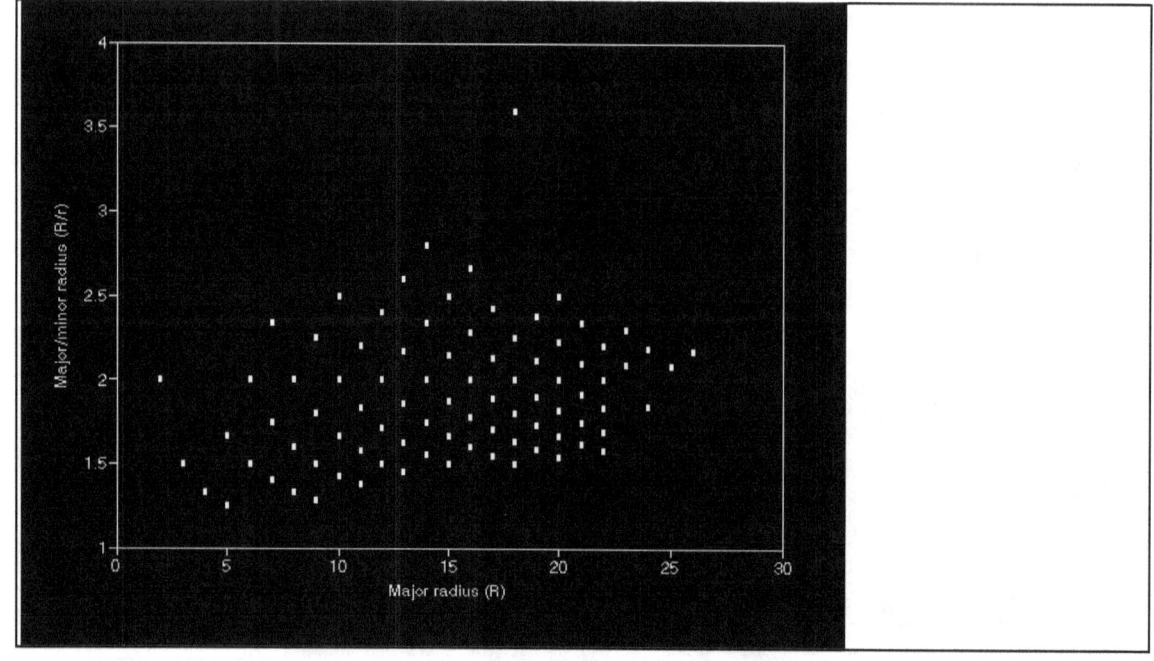

Table 5.10 *Echinaster luzonicus*

The mean (R mm and R/r), standard deviation (R mm and R/r) and sample size (N) for each sampling period and a grand mean of both R (mm) and R/r, and total sample size are tabled.

PERIOD	MEAN(R)	S.D.(R)	MEAN(R/r)	S.D.(R/r)	N
MAY 1978	40.95	10.88	5.57	0.86	112
AUG 1978	48.19	16.77	-	-	73
NOV 1978	40.29	10.90	-	-	21
JUN 1979	52.08	11.84	6.10	0.98	38
SEP 1979	51.56	11.66	6.01	0.92	106
DEC 1979	54.54	16.65	6.09	0.97	67
APR 1980	43.72	8.39	5.45	0.75	83
JUL 1980	45.90	12.11	5.55	0.96	110
NOV 1980	46.43	11.23	5.90	0.78	81
JUL 1981	45.78	16.32	6.26	1.13	65
MAY 1982	50.74	15.36	5.99	1.28	76
OCT 1982	60.67	17.79	6.27	1.35	30
DEC 1982	47.28	11.00	5.80	1.00	126
TOTAL	47.66		5.84		988

********** ANOVA **********

Variation in mean Major Radius (R) with respect to time.

Source of Variation	df	SS	MS	F
Among	10.0	18342.5	1834.3	11.4
Within	883.0	141795.3	160.6	
Total	893.0	160137.8		

$p < .001$

Figure 5.10a *Echinaster luzonicus*
Frequency distribution of major radius (R mm).

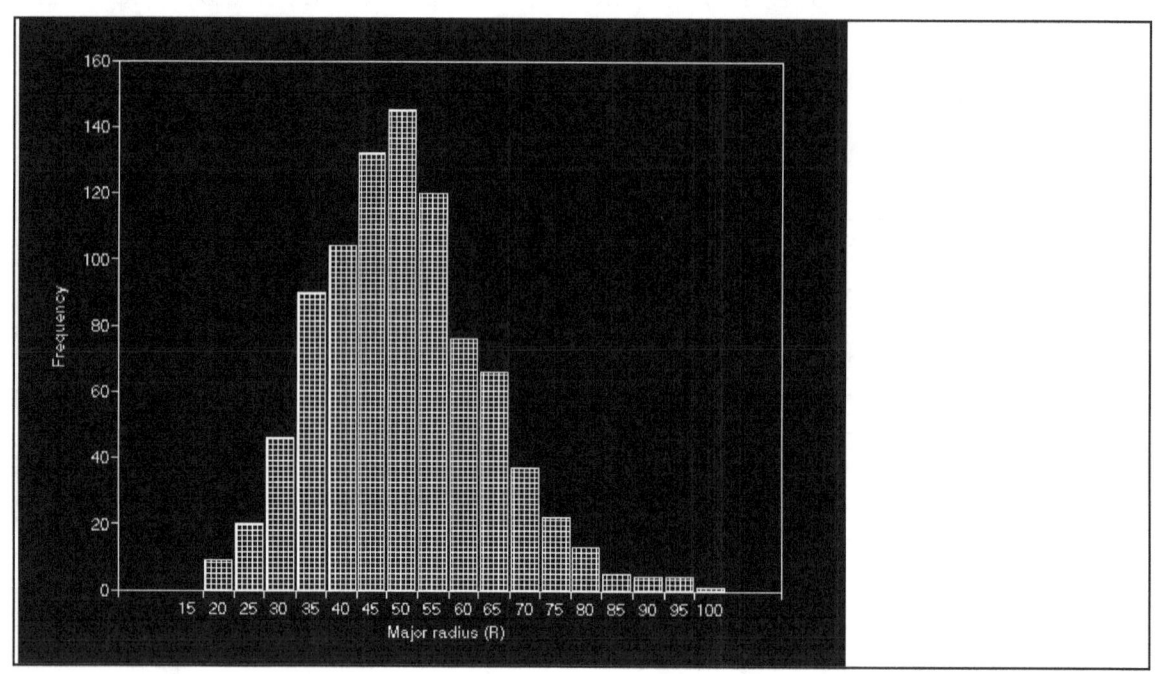

Figure 5.10b *Echinaster luzonicus*
Frequency distribution of major radius / minor radius (R/r)

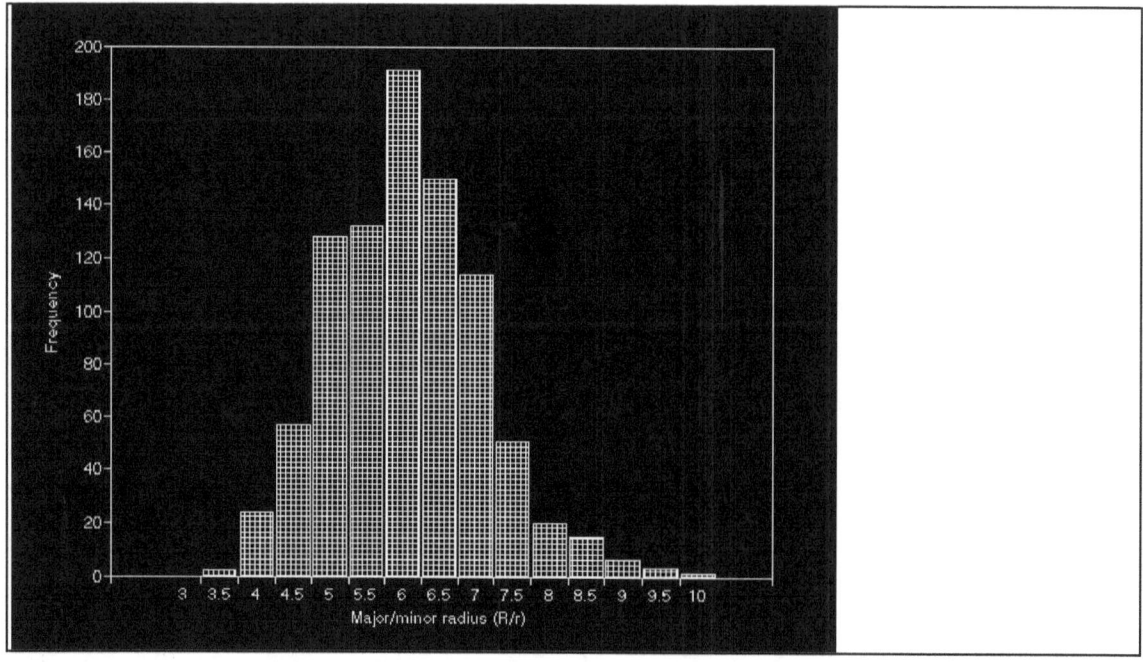

Figure 5.10c *Echinaster luzonicus*
Relation between minor radius (r mm) and major radius (R mm)

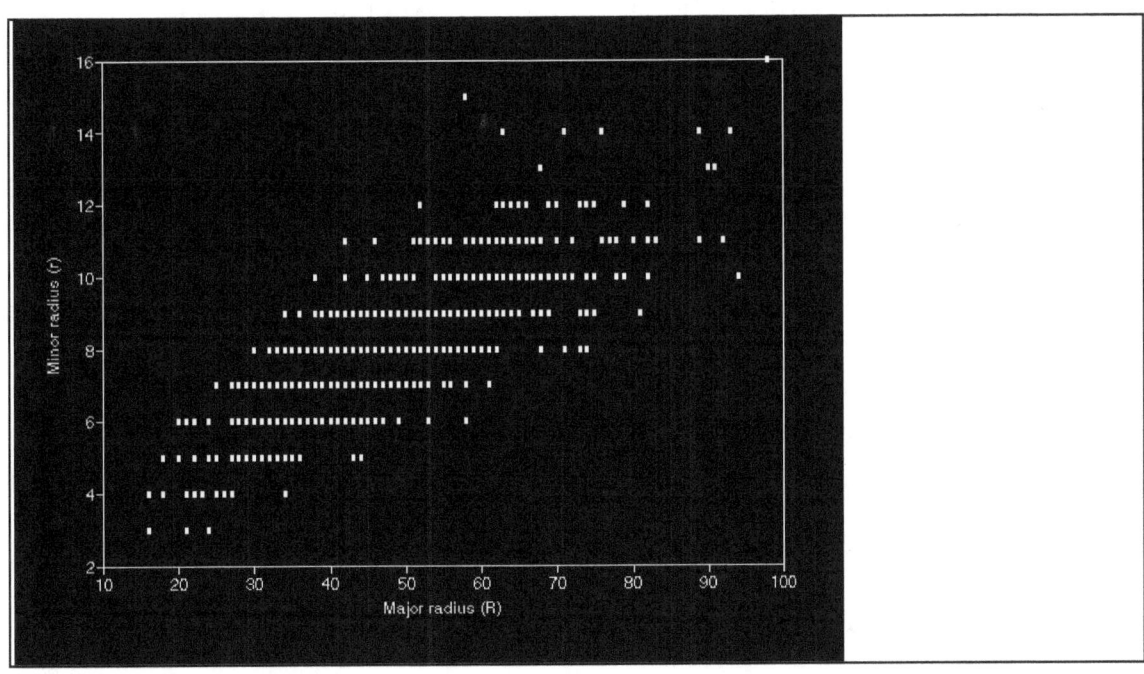

Figure 5.10d *Echinaster luzonicus*
Relation between major radius / minor radius (R/r)
and major radius (R mm)

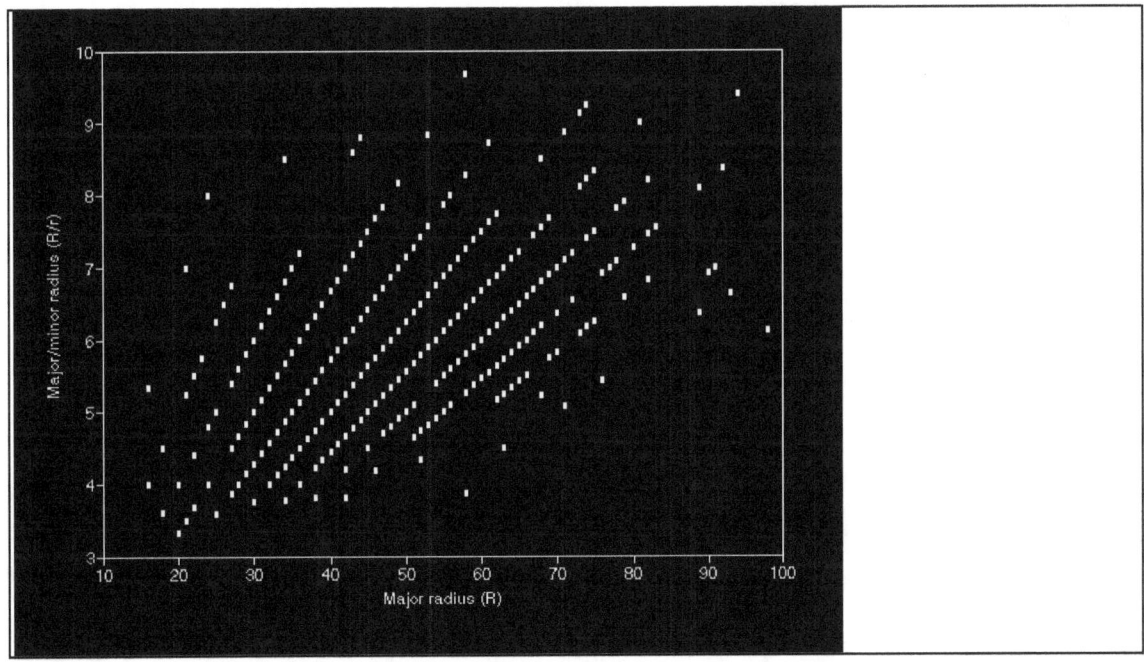

Table 5.11

The mean major radius (MEAN R mm), mean major radius / minor radius (MEAN R/r) and sample size (N) of the species that occurred on traverses at Heron Reef.

	MEAN R mm	MEAN R/r	N
Culcita novaeguineae	109	1.3	12
Asteropsis carinifera	74	2.8	3
Dactylosaster cylindricus	83	-	1
Fromia elegans	32	3.9	183
Fromia milleporella	30	2.3	1
Gomophia egyptiaca	45	4.8	10
Linckia guildingii	134	9.8	131
Linckia laevigata	127	6.5	516
Linckia multifora	38	7.3	396
Nardoa novaecaledoniae	88	5.7	361
Nardoa pauciforis	104	6.6	233
Nardoa rosea	88	6.2	7
Ophidiaster armatus	55	6.3	7
Ophidiaster confertus	86	8.2	4
Ophidiaster granifer	27	4.3	121
Ophidiaster lioderma	105	-	1
Ophidiaster robillardi	37	6.9	21
Asterina anomala	4	1.6	16
Asterina burtoni	13	1.8	203
Disasterina abnormalis	15	1.9	1109
Disasterina leptalacantha	13	2.0	8
Tegulaster emburyi	18	2.3	1
Echinaster luzonicus	48	5.8	988
Coscinasterias calamaria	19	4.4	8

5.4 Discussion

The species of starfish that occurred on the traverses can be grouped according to general body size. The species that had a maximum major radius (R) greater than 100 mm were regarded as large-bodied. The large-bodied species (maximum major radius in parenthesis) are *Culcita novaeguineae* (130 mm), *Linckia guildingii* (240 mm), *Linckia laevigata* (190 mm), *Nardoa novaecaledoniae* (115 mm), *Nardoa pauciforis* (135 mm), *Nardoa rosea* (101 mm), *Ophidiaster confertus* (102 mm) and *Ophidiaster lioderma* (105 mm).

Small-bodied species had a maximum major radius (R) that was less than 100mm. These species were *Asteropsis carinifera* (85 mm), *Dactylosaster cylindricus* (83 mm), *Fromia elegans* (46 mm), *Fromia milleporella* (30 mm), *Gomophia egyptiaca* (64 mm), *Linckia multifora* (90 mm), *Ophidiaster armatus* (73 mm), *Ophidiaster granifer* (48 mm), *Ophidiaster robillardi* (50 mm), *Asterina anomala* (7 mm), *Asterina burtoni* (23 mm), *Disasterina abnormalis* (27 mm), *Disasterina leptalacantha* (20 mm), *Tegulaster emburyi* (18 mm), *Echinaster luzonicus* (100 mm) and *Coscinasterias calamaria* (30 mm).

With the exception of *Ophidiaster confertus* and *Ophidiaster lioderma*, large-bodied species possessed an extremely tough body wall. *Ophidiaster confertus* is a temperate species and *Ophidiaster lioderma* (an extremely rare species) is covered with a greatly thickened skin. The cut-off distinguishing large-bodied from small-bodied starfish at maximum R = 100 mm is arbitrary, and both *Asteropsis carinifera* and *Dactylosaster cylindricus* could be included in this large-bodied group if the distinction was based on mean size. The mean size of the two next largest small-bodied species, *Linckia multifora* and *Echinaster luzonicus*, remained about half the maximum size through a continuing process of autotomy.

It can be seen from the data presented in Tables 5.1 to 5.11 and Figures 5.1a to 5.10a that juveniles of the relatively common, sexually reproducing, large-bodied asteroids, *Linckia laevigata*, *Nardoa novaecaledoniae* and *Nardoa pauciforis* were rare and the populations of these species were adult dominated throughout the study period. Relatively small specimens of *Linckia guildingii* resulting from asexual reproduction were observed but these individuals represent only a minor component of the adult dominated population. In all large bodied species, distinct year classes were not observed in the population size structures. While a highly variable growth rate can disguise a dominant year class, the fact that neither numerous small individuals (indicating high recruitment) nor obvious population declines (evidence of high mortality) was observed over a period of several years, suggests strongly that these species are long-lived (persistent).

Small individuals were more common in the populations of *Linckia multifora*, *Echinaster luzonicus* and *Disasterina abnormalis*, and to a lesser extent in the populations of *Ophidiaster granifer* and *Asterina burtoni*. These juveniles resulted from either sexual or asexual reproduction. While small specimens occurred in the sub-tidal population of *Fromia elegans*, its population structure was still adult dominated throughout the period of study.

Many hypotheses have been proposed to explain the apparent paucity of juveniles amongst coral-reef echinoderms. Juveniles may occupy such different habitats from the adults that they have not been adequately sampled or the adult animals may be long lived and recruitment low (Yamaguchi, 1977 a). Recruitment may also be patchy in both time and space (Yamaguchi, 1973 a, 1973 b, 1977 b). In studies of some fishes, it has been shown that each reproductive season large quantities of sperm and eggs are released but owing to the rigours of planktonic life and the uncertainty of locating settling substrate, most larvae are lost before settlement (Sale, 1976, 1977). Endean and Cameron (1990 b) proposed that the mortality that regulates the adult population density of *Acanthaster planci* occurs on post-settlement stages.

The giant triton (*Charonia tritonis*) and other members of the genus *Charonia* are known predators of many species of starfish (Chesher, 1969 b; Endean, 1969; Laxton, 1971; Noguchi et al., 1982; Percharde, 1972) and other predators of starfish include shrimp (Glynn, 1974), a worm (Glynn, 1984), fish (Ormond et al., 1973) and other starfish (Mauzey et al, 1968; Dayton et al, 1977; Birkeland et al, 1982). In most species at Heron Island, evidence of starfish mortality was hard to find. This has also been true for *Acanthaster planci*, even following population outbreaks. A high incidence of sub-lethal predation on adult *Acanthaster planci* was reported by McCallum et al. (1989), who suggested that lethal predation could account for the paucity of juveniles in populations of *Acanthaster planci*.

Parasitism of starfish by molluscs is well known (see e.g. Davis, 1967; Elder, 1979; Egloff et al., 1988). At Heron Reef, the incidence of infection by molluscs was low, except in *Linckia multifora* and *Ophidiaster granifer*. Bouillon and Jangoux (1985) recorded a high proportion of *Linckia laevigata* infected, but a high rate of infection of this species did not occur at Heron Reef.

Yamaguchi (1977 a) showed a much higher abundance of *Linckia laevigata* at Guam (a reef known to carry *A. planci* outbreaks) but the mean individual size was much smaller than in the present study. Thompson and Thompson (1982) and Laxton (1974) also found a smaller mean size of *Linckia laevigata* compared with that found in this study. Thompson and Thompson's study was conducted at Lizard Island, Queensland (a reef known to carry *A. planci* outbreaks) and a greater spatial variation in density and mean size was found than was evident at Heron Reef. While Laxton's samples were taken from Heron Reef in approximately the same habitat as was used in this study, the abundance of *Linckia laevigata* is not stated and there appears to have been confusion with *Linckia multifora*. Observations made at Lady Musgrave Reef in the Bunker Group where *Linckia laevigata* was much more abundant than at Heron Reef also show a smaller mean size of specimens from some habitats. Lady Musgrave Reef is known to have carried a minor outbreak of *Acanthaster planci*. The data indicate that abundance and mean individual size may be inversely correlated, but abundance appears to be more closely regulated on a reef such as Heron Reef that is not known to have carried an *A. planci* outbreak. Apart from the destruction of the hard coral cover caused by such outbreaks, great changes occur in the fauna and flora of reefs following *A. planci* population outbreaks (Endean and Cameron, 1990 b).

CHAPTER 6

SEXUAL REPRODUCTION

6.1 Introduction

Most asteroids possess 10 gonads, two in each ray with gonoducts opening in the interradii. In some coral-reef genera, such as *Linckia* and *Nardoa*, the gonads are arranged serially with numerous gonoducts existing along the length of the arm. When an individual is ready to spawn, the gonads can occupy the whole length of a ray. In most species, the sexes are separate, but hermaphroditism has been reported in *Asterina burtoni* (Achituv, 1972; Achituv and Malik, 1985; Achituv and Sher, 1991). In all coral-reef species studied previously, fertilisation is external with gametes being released directly into the water. An off-reef species, *Euretaster insignis*, belongs to the family Pterasteridae, other members of which are known to brood their young within the supra-dorsal membrane (McClary and Mladenov, 1989; McClary, 1990). The spawning of one individual might trigger other individuals to spawn, thus increasing the chance of fertilisation (Okaji, 1991). Alternately, synchronous spawning might be triggered extrinsically (see Yamaguchi and Lucas, 1984; Minchin, 1987).

The sexual reproductive cycle has been studied in some of the commoner coral-reef species e.g. *Asterina burtoni*, (by Achituv, 1972; James, 1972; Achituv and Malik, 1985), *Linckia laevigata* (by Yamaguchi, 1977 a), and *Ophidiaster granifer* (by Yamaguchi and Lucas, 1984). The type of larval development has also been studied in several species e.g. *Astropecten polyacanthus* (by Oguro et al., 1975), *Acanthaster planci* (by e.g. Henderson and Lucas, 1971), *Gomophia egyptiaca* (by Yamaguchi, 1974), *Leiaster leachi* (by Komatsu, 1973), *Ophidiaster granifer*, *Ophidiaster robillardi* and *Ophidiaster squamous* (by Yamaguchi and Lucas, 1984). The known forms of reproduction along with the type of larval development of most Guam species are tabled by Yamaguchi (1975 b).

The use of the hormone 1-methyl adenine to produce final maturation and subsequent release of gametes in asteroids is well documented (Kanatani, 1969, 1973). Yamaguchi (1977 a) described the injection of the hormone into the coelomic cavity of *Linckia laevigata* to assess the stage of development of the gonads. This procedure was also used by Yamaguchi and Lucas (1984). The presence and strength of response to treatment have been shown to depend upon the stage of maturation of the gametocytes (Kuborta *et al.*, 1977).

The time required before a response is produced, following treatment, depends upon the proximity of the natural breeding season (Kanatani, 1969). A delayed response in the genus *Echinaster* was described previously by Turner (1976). Additionally, it is possible that not all individuals within a population are at exactly the same stage of gamete development at any time (see Pearse, 1968).

6.2 Methods

The reproductive analysis of each of the species entailed the injection of 1-methyl adenine into the arms of a sample of the population several times during each year. The chemical was obtained as anhydrous powder in 10 mg tubes from Sigma Chemicals. The anhydrous powder was kept frozen following delivery. The concentration of 1-methyl adenine used for injection was 0.0001 M. dissolved in sea-water. A working solution was prepared freshly for each sampling period and stored in a refrigerator at 4°C. It was necessary to warm the sea-water temporarily to 40°C to facilitate the dissolution of the chemical as the working solution was being prepared. Once the chemical was dissolved (requiring about five to ten minutes with stirring) it was immediately placed in the refrigerator. The solution was not warmed again before injection into the starfish.

The quantity injected per individual depended on the mean size of the species tested. It ranged from one millilitre per individual in small species such as *Ophidiaster granifer* to five millilitres per individual in large species such as *Linckia laevigata*. In large species, the hormone was injected into three of the arms of each animal while it was temporally removed from the aquarium. In smaller species the injection was administered aborally into one interradius. Following injection, each animal was returned to its aquarium along with other conspecifics which had also been injected. The test animals were then observed for several hours and any release of gametes through the gonopores was recorded. To maintain visibility within the test aquaria, specimens were removed and placed in a larger tank once they commenced spawning. Spawning always continued following the transfer. Gametes that were fertilised in aquaria were never released in the field.

The procedure used by Kanatani (1969) required the extraction of gonad for *in vitro* treatment with 1-methyl adenine. Other methods of determining reproductive periodicity, such as gonad index or histological examination require the test individuals to be killed, and the gonads removed. Many species of coral-reef asteroid occur in low abundance and the regular killing of test individuals would have required the use of much smaller sample sizes to ensure that the population was not reduced by periodic testing. In the present study the convenience of an *in vivo* treatment, that allowed the rapid testing of a large number of individuals of each species, was considered to outweigh the limitations imposed by the lack of detailed histological information.

Possible variability in response within samples required sample sizes in the vicinity of 20 to 30 individuals to ensure statistical significance of the different spawning frequencies, at different times of the year. Histological study would have provided direct qualitative evidence of gametogenesis. The dichotomous (presence / absence) spawning data obtained in the present study required larger sample sizes to demonstrate any periodicity conclusively. The G Test was used to establish that the observed response in the breeding season was significantly different from the null (low all year round) response. While in some cases, the "expecteds" were as low as two, the G statistic appeared sufficiently high to indicate a significant spawning response.

6.3 Results

Release of gametes required no longer than 3 hours after injection except in *Echinaster luzonicus*. However, for every species that was studied, the response time varied throughout the year. This time ranged from just under three hours, two months before the breeding season, to as little as 15 minutes at the peak of the season. At this peak, if the water temperature within aquaria was allowed to rise above that of the reef flat (as it would on a very hot day), spontaneous spawning was observed in species of *Linckia* and *Nardoa*. Spawning in the field was not observed during this study. The result with 1-methyl adenine was always reduced if the test animals had undergone previous spawning.

Tables 6.1a to 6.8a list the spawning response to injection with 1-methyl adenine and Figures 6.1 to 6.8 graph the annual spawning pattern of each of the common species over the study period. Tables 6.1b to 6.8b show the results of the G test, comparing the spawning response in four seasons. Table 6.9 lists the reproductive strategies of each species.

Providing that the water did not become too cloudy, it was always possible to determine the sex of the individuals by the type of gamete released. The size, number and development of eggs was not studied in detail, but varied among species depending on the type of larvae produced.

Table 6.1a *Fromia elegans*

The spawning response to injection with 1-methyl adenine (1.M.A.) at different sampling periods. The sample size (N), quantity injected (ml), numbers of both male and female starfish and percent of sample which spawned are tabled.

PERIOD	N	1.M.A.	MALES	FEMALES	%
MAY 1978	-	-	-	-	-
AUG 1978	-	-	-	-	-
NOV 1978	6	1.0	5	1	100
FEB 1979	24	1.0	0	0	0
JUN 1979	10	1.0	0	0	0
SEP 1979	7	1.0	0	0	0
DEC 1979	14	1.0	1	0	7
APR 1980	20	1.0	0	0	0
JUL 1980	20	1.0	0	0	0
NOV 1980	20	1.0	0	0	0
JUL 1981	20	1.0	0	0	0
JAN 1982	30	0.5	0	0	0
MAY 1982	20	1.0	0	0	0
OCT 1982	30	1.0	0	0	0
DEC 1982	18	1.0	1	2	16
TOTAL	239		7	3	4

Figure 6.1 *Fromia elegans*

Annual Reproductive Cycle

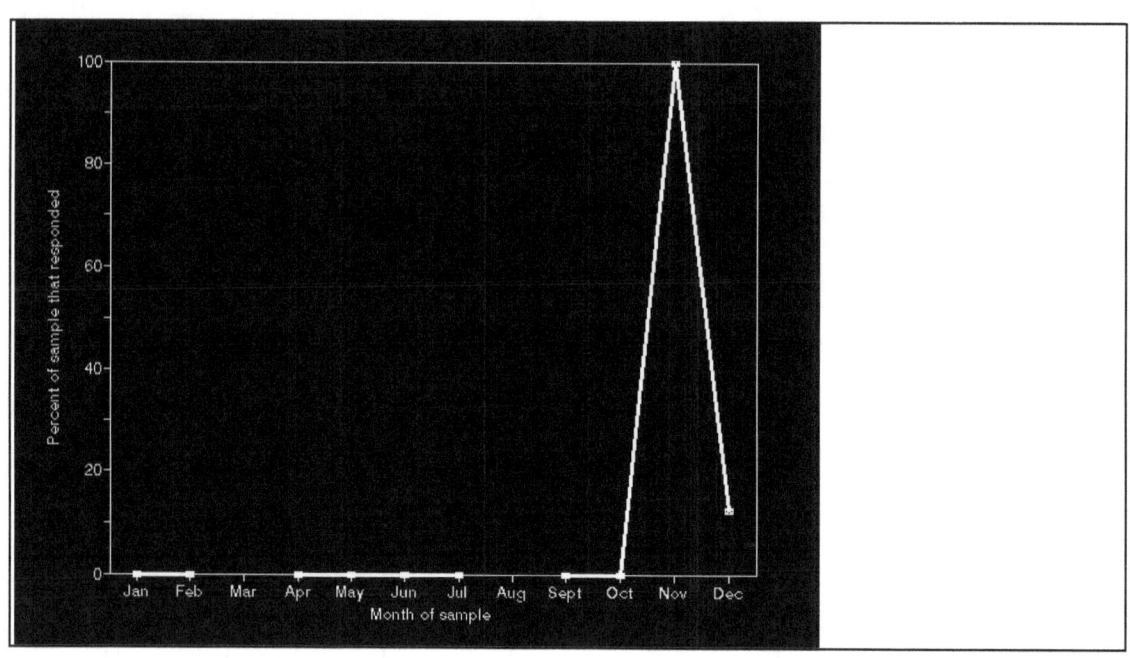

Table 6.1b *Fromia elegans*

G test for independence in a 4 X 2 table.
Spawning frequency of four periods by spawning response

Season	Not spawning	Spawning	Σ
Feb - Apr	54	0	54
May - Jul	50	0	50
Aug - Oct	47	0	47
Nov - Jan	78	10	88
Σ	229	10	239
G = 20.740	d.f. = 3	P < 0.001	

Table 6.2a *Linckia guildingii*

The spawning response to injection with 1-methyl adenine (1.M.A.) at different sampling periods. The sample size (N), quantity injected (ml), numbers of both male and female starfish and percent of sample which spawned are tabled.

PERIOD	N	1.M.A.	MALES	FEMALES	%
AUG 1978	10	5.0	0	0	0
NOV 1978	10	5.0	0	0	0
FEB 1979	8	5.0	0	0	0
SEP 1979	10	5.0	0	0	0
DEC 1979	6	5.0	0	2	33
APR 1980	8	5.0	0	0	0
JUL 1980	6	5.0	0	0	0
NOV 1980	6	5.0	0	0	0
JAN 1982	17	2.5	3	1	23
MAY 1982	3	2.5	0	0	0
OCT 1982	4	2.5	0	0	0
DEC 1982	8	2.5	4	0	50
TOTAL	96		7	3	10

Figure 6.2 *Linckia guildingii*

Annual Reproductive Cycle

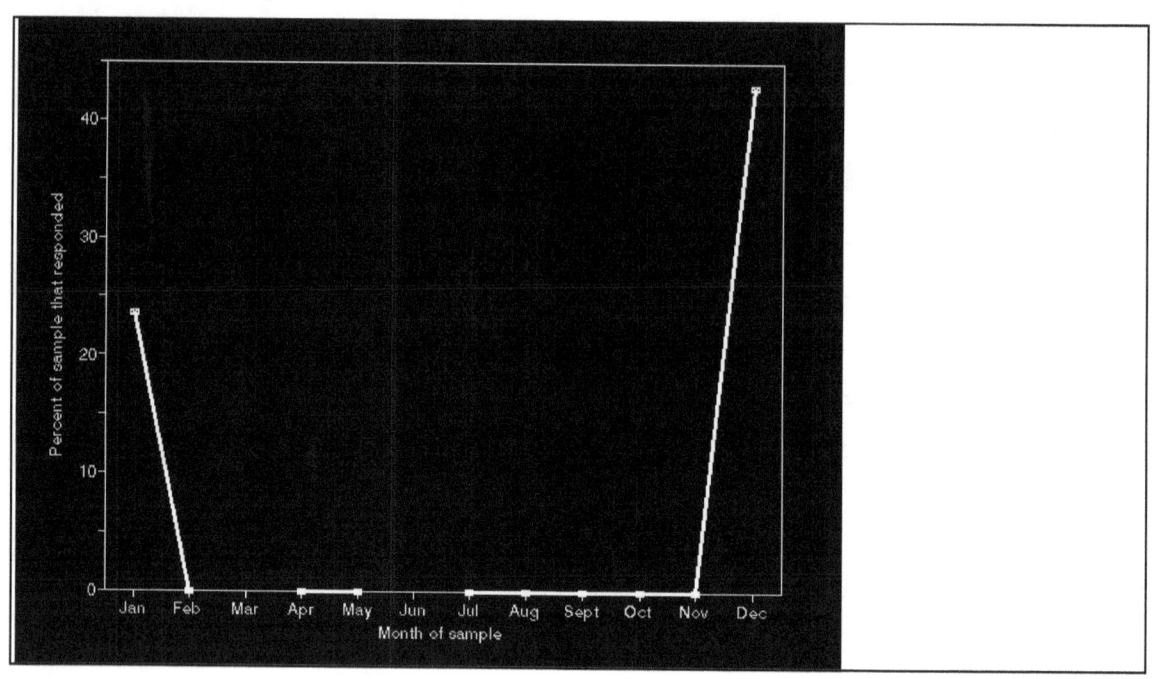

Table 6.2b *Linckia guildingii*

G test for independence in a 4 X 2 table.
Spawning frequency of four periods by spawning response

Season	Not spawning	Spawning	Σ
Feb - Apr	26	0	26
May - Jul	9	0	9
Aug - Oct	14	0	14
Nov - Jan	37	10	47
Σ	86	10	96
G = 15.501	d.f. = 3	P < 0.005	

Table 6.3a *Linckia laevigata*

The spawning response to injection with 1-methyl adenine (1.M.A.) at different sampling periods. The sample size (N), quantity injected (ml), numbers of both male and female starfish and percent of sample which spawned are tabled.

PERIOD	N	1.M.A.	MALES	FEMALES	%
MAY 1978	3	5.0	0	0	0
AUG 1978	10	5.0	6	2	80
NOV 1978	10	5.0	3	3	60*
FEB 1979	20	5.0	0	0	0
JUN 1979	26	5.0	3	0	11
SEP 1979	26	5.0	12	8	76
DEC 1979	20	5.0	8	5	65*
APR 1980	40	5.0	1	0	2
JUL 1980	40	5.0	**	**	67
NOV 1980	20	5.0	11	5	80*
JUL 1981	40	5.0	25	11	90
JAN 1982	30	5.0	0	1	3
MAY 1982	30	5.0	0	0	0
OCT 1982	26	5.0	12	13	96
DEC 1982	20	5.0	0	2	10
TOTAL	361		81	50	36

* prior spontaneous spawning

** 27 in total (sex not determined)

Figure 6.3 *Linckia laevigata*

Annual Reproductive Cycle

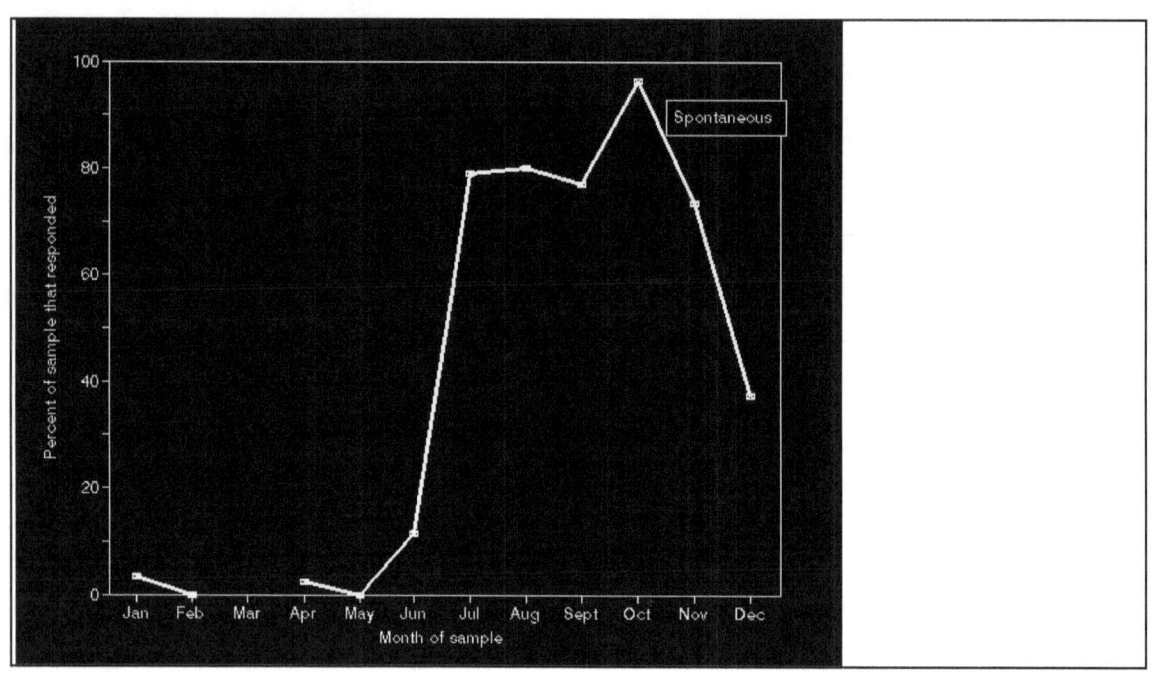

Table 6.3b *Linckia laevigata*

G test for independence in a 4 X 2 table.
Spawning frequency of four periods by spawning response

Season	Not spawning	Spawning	Σ
Jan - Mar	49	1	50
Apr - Jun	95	4	99
Jul - Sep	25	91	116
Oct - Dec	34	62	96
Σ	203	158	361
G = 205.807	d.f. = 3	P < 0.001	

Table 6.4a *Nardoa novaecaledoniae*

The spawning response to injection with 1-methyl adenine (1.M.A.) at different sampling periods. The sample size (N), quantity injected (ml), numbers of both male and female starfish and percent of sample which spawned are tabled.

PERIOD	N	1.M.A.	MALES	FEMALES	%
MAY 1978	10	2.5	0	0	0
AUG 1978	10	2.5	0	0	0
NOV 1978	10	2.5	0	0	0
FEB 1979	20	2.5	5	3	40
JUN 1979	16	2.5	0	0	0
SEP 1979	21	2.5	0	0	0
DEC 1979	25	2.5	0	0	0
APR 1980	30	2.5	0	0	0
JUL 1980	30	2.5	0	0	0
NOV 1980	20	2.5	0	0	0
JUL 1981	30	2.5	0	0	0
JAN 1982	18	2.5	5	8	72
MAY 1982	20	2.5	0	0	0
OCT 1982	27	2.5	0	0	0
DEC 1982	20	2.5	1	0	5
TOTAL	307		11	11	7

Figure 6.4 *Nardoa novaecaledoniae*

Annual Reproductive Cycle

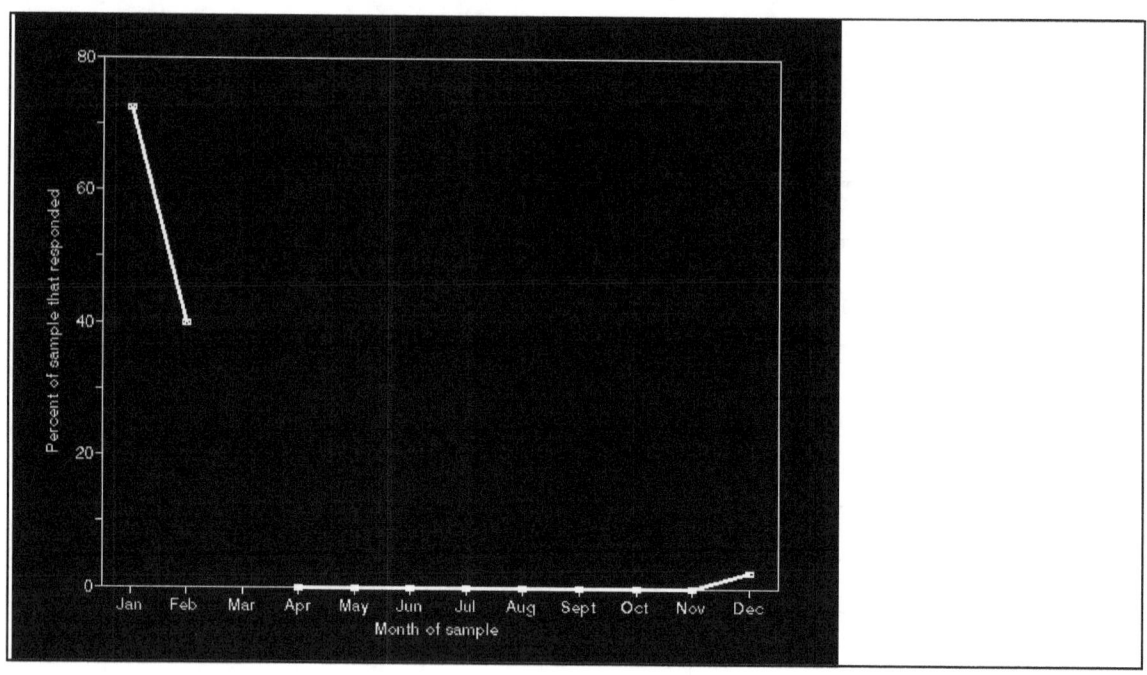

Table 6.4b *Nardoa novaecaledoniae*

G test for independence in a 4 X 2 table.
Spawning frequency of four periods by spawning response

Season	Not spawning	Spawning	Σ
Jan - Mar	17	21	38
Apr - Jun	76	0	76
Jul - Sep	91	0	91
Oct - Dec	101	1	102
Σ	285	22	307
G = 94.862	d.f. = 3	P < 0.001	

Table 6.5a *Nardoa pauciforis*

The spawning response to injection with 1-methyl adenine (1.M.A.) at different sampling periods. The sample size (N), quantity injected (ml), numbers of both male and female starfish and percent of sample which spawned are tabled.

PERIOD	N	1.M.A.	MALES	FEMALES	%
MAY 1978	10	2.5	0	0	0
AUG 1978	10	2.5	0	0	0
NOV 1978	14	2.5	6	1	50
FEB 1979	20	2.5	0	0	0
JUN 1979	10	2.5	0	0	0
SEP 1979	13	2.5	0	0	0
DEC 1979	14	2.5	8	5	92
APR 1980	20	2.5	0	0	0
JUL 1980	18	2.5	0	0	0
NOV 1980	15	2.5	0	0	0
JUL 1981	25	2.5	0	0	0
JAN 1982	17	2.5	7	0	41
MAY 1982	11	2.5	0	0	0
OCT 1982	15	2.5	0	0	0
DEC 1982	19	2.5	12	0	63*
TOTAL	231		33	6	39

* prior spontaneous spawning

Figure 6.5 *Nardoa pauciforis*

Annual Reproductive Cycle

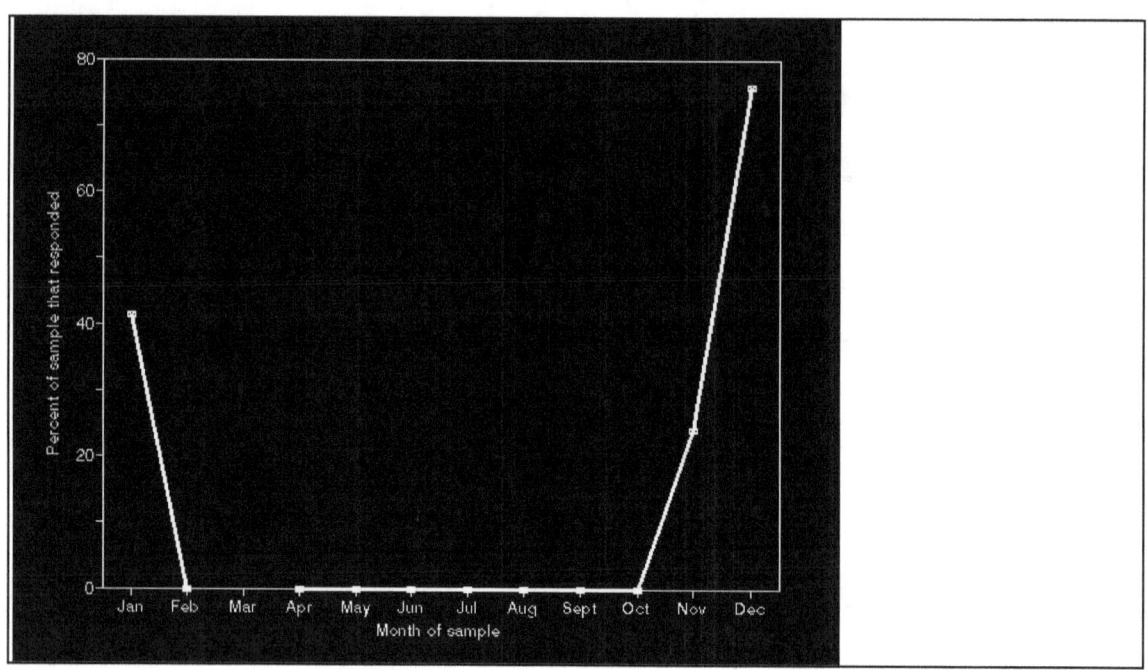

Table 6.5b *Nardoa pauciforis*

G test for independence in a 4 X 2 table.
Spawning frequency of four periods by spawning response

Season	Not spawning	Spawning	Σ
Feb - Apr	40	0	40
May - Jul	74	0	74
Aug - Oct	38	0	38
Nov - Jan	40	39	79
Σ	192	39	231
G = 100.256	d.f. = 3	P < 0.001	

Table 6.6a *Ophidiaster granifer*

The spawning response to injection with 1-methyl adenine (1.M.A.) at different sampling periods. The sample size (N), quantity injected (ml), numbers of both male and female starfish and percent of sample which spawned are tabled.

PERIOD	N	1.M.A.	MALES	FEMALES	%
MAY 1978	10	1.0	0	0	0
AUG 1978	10	1.0	0	0	0
NOV 1978	8	1.0	0	4	50
FEB 1979	9	1.0	0	0	0
JUN 1979	6	1.0	0	0	0
DEC 1979	3	1.0	0	0	0
APR 1980	4	1.0	0	0	0
JUL 1980	6	1.0	0	0	0
NOV 1980	1	1.0	0	0	0
JUL 1981	6	1.0	0	0	0
JAN 1982	10	0.5	0	0	0
MAY 1982	8	1.0	0	0	0
OCT 1982	12	1.0	0	0	0
DEC 1982	11	1.0	0	3	27
TOTAL	104		0	7	7

Figure 6.6 *Ophidiaster granifer*

Annual Reproductive Cycle

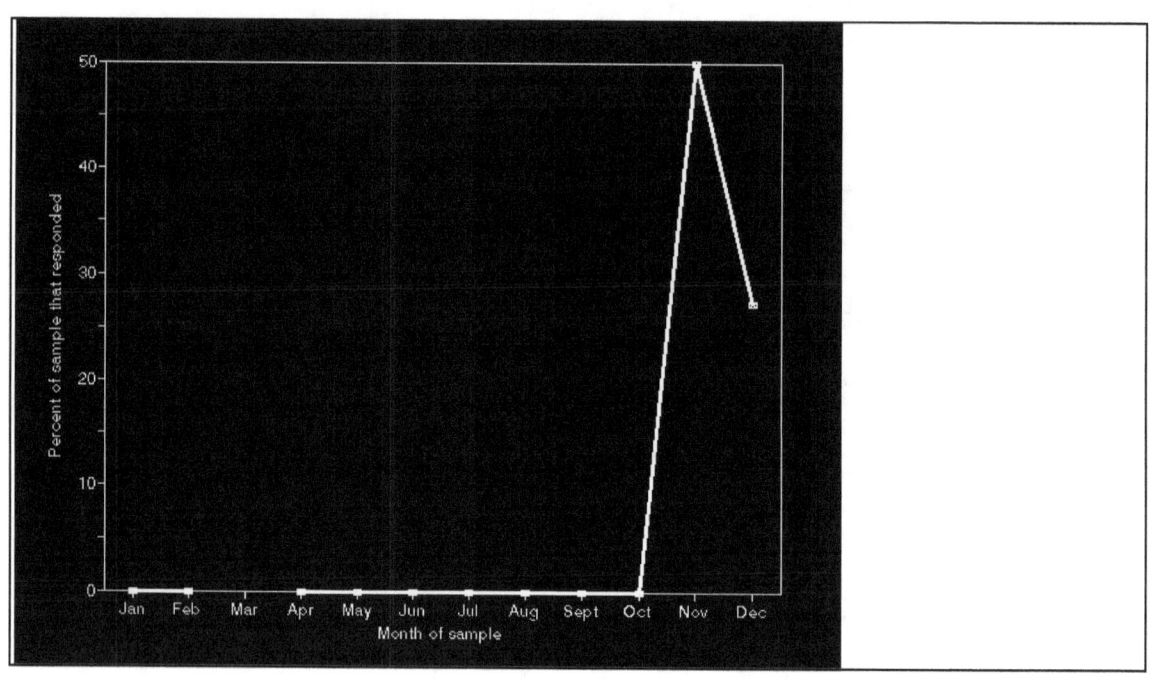

Table 6.6b *Ophidiaster granifer*

G test for independence in a 4 X 2 table.
Spawning frequency of four periods by spawning response

Season	Not spawning	Spawning	Σ
Feb - Apr	13	0	13
May - Jul	36	0	36
Aug - Oct	22	0	22
Nov - Jan	26	7	33
Σ	97	7	104
G = 17.191	d.f. = 3	P < 0.001	

Table 6.7a *Disasterina abnormalis*

The spawning response to injection with 1-methyl adenine (1.M.A.) at different sampling periods. The sample size (N), quantity injected (ml), numbers of both male and female starfish and percent of sample which spawned are tabled.

PERIOD	N	1.M.A.	MALES	FEMALES	%
MAY 1978	10	1.0	0	0	0
AUG 1978	10	1.0	0	0	0
NOV 1978	10	1.0	0	1	10
FEB 1979	10	1.0	0	0	0
JUN 1979	10	1.0	0	0	0
SEP 1979	20	1.0	4	3	35
DEC 1979	20	1.0	0	0	0
APR 1980	19	1.0	0	0	0
JUL 1980	30	1.0	3	0	10
NOV 1980	20	1.0	0	0	0
JUL 1981	10	1.0	1	0	10
JAN 1982	20	0.5	0	0	0
MAY 1982	20	1.0	0	0	0
OCT 1982	30	1.0	6	8	46
DEC 1982	10	1.0	0	0	0
TOTAL	249		11	12	9

Figure 6.7 *Disasterina abnormalis*

Annual Reproductive Cycle

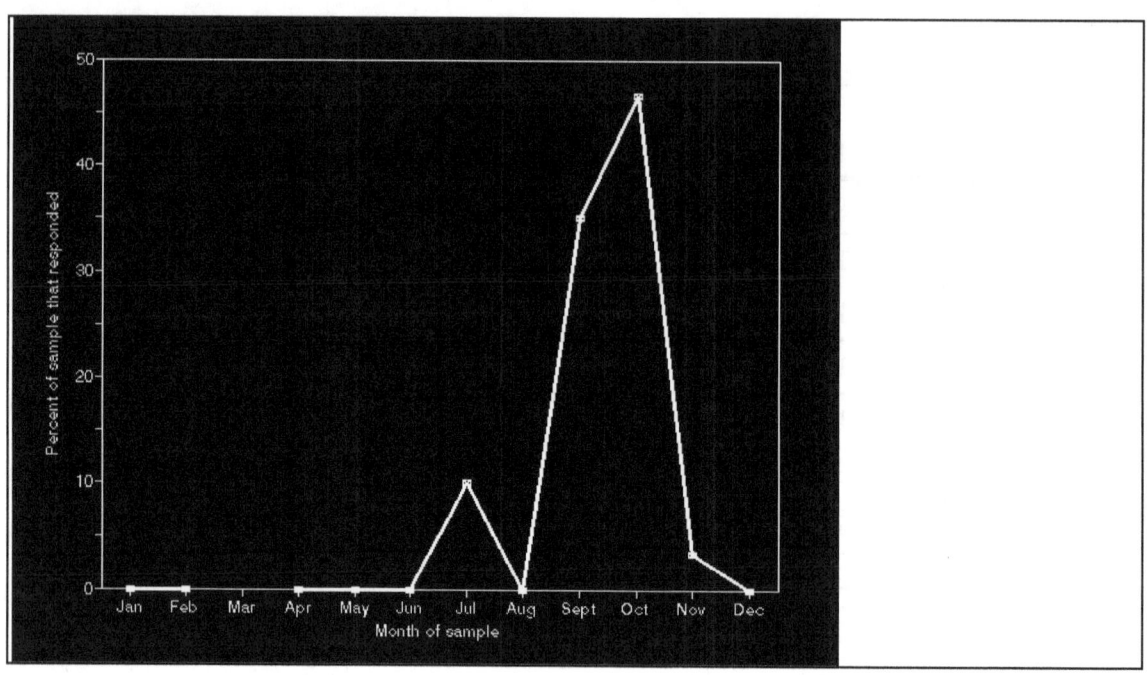

Table 6.7b *Disasterina abnormalis*

G test for independence in a 4 X 2 table.
Spawning frequency of four periods by spawning response

Season	Not spawning	Spawning	Σ
Jan - Mar	30	0	30
Apr - Jun	59	0	59
Jul - Sep	59	11	70
Oct - Dec	75	15	90
Σ	223	26	249
G = 24.685	d.f. = 3	P < 0.001	

Table 6.8a *Echinaster luzonicus*

The spawning response to injection with 1-methyl adenine (1.M.A.) at different sampling periods. The sample size (N), quantity injected (ml), numbers of both male and female starfish and percent of sample which spawned are tabled.

PERIOD	N	1.M.A.	MALES	FEMALES	%
MAY 1978	10	2.5	0	0	0
AUG 1978	10	2.5	0	0	0
NOV 1978	10	2.5	0	0	0
FEB 1979	20	2.5	3	1	20
JUN 1979	20	2.5	0	0	0
SEP 1979	20	2.5	0	0	0
DEC 1979	20	2.5	2	0	10
APR 1980	20	2.5	0	1	5
JUL 1980	20	2.5	0	0	0
NOV 1980	20	2.5	0	0	0
JUL 1981	20	2.5	0	0	0
JAN 1982	20	1.0	0	0	0
MAY 1982	30	2.5	0	0	0
OCT 1982	30	2.5	0	0	0
DEC 1982	20	2.5	0	0	0
TOTAL	290		5	2	2

Figure 6.8 *Echinaster luzonicus*

Annual Reproductive Cycle

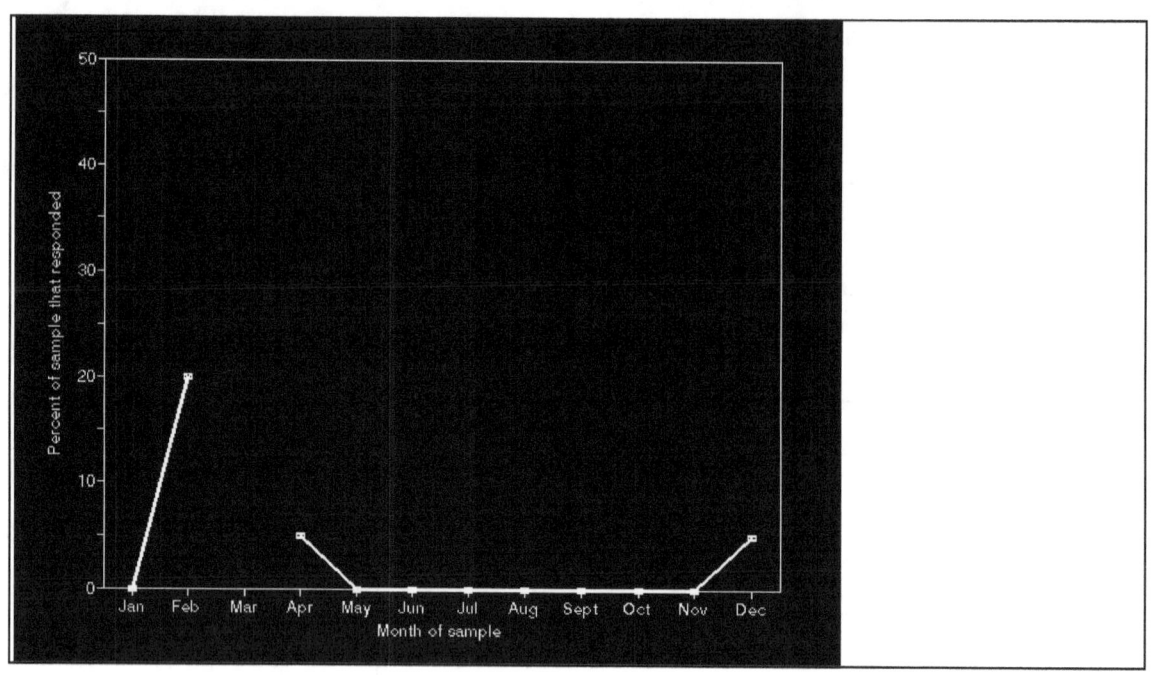

Table 6.8b *Echinaster luzonicus*

G test for independence in a 4 X 2 table.
Spawning frequency of four periods by spawning response

Season	Not spawning	Spawning	Σ
Feb - Apr	35	5	40
May - Jul	100	0	100
Aug - Oct	60	0	60
Nov - Jan	88	2	90
Σ	283	7	290
G = 16.642	d.f. = 3	$P < 0.001$	

The results of observations, relating to 1-methyl adenine injections, on the sexual reproductive cycles of the commoner asteroid species at Heron Reef are outlined below.

Females of *Fromia elegans* release from 100 to 200 very large (approximately 2.0 mm diameter) eggs from gonopores, two of which are located in each interradius. The eggs are bright red in colour and are of neutral buoyancy. Egg release may take up to three hours following injection but can occur after only 30 minutes. The males spawn within one hour and sperm are released through gonopores which are located slightly higher in each interradius than in the female. The fertilised eggs undergo lecithotrophic development. The peak of sexual activity occurred in early summer (November, December).

Females of *Linckia guildingii* release large numbers of small (approximately 0.1 mm diameter), colourless and negatively buoyant eggs through gonopores located serially along the arms. At no time was the sexually active proportion of the population very high. The peak in sexual activity that was apparent occurred in mid summer (December).

The reproductive cycle of *Linckia laevigata*, at Guam, was studied by Yamaguchi (1977 a). Females release very large numbers of small (approximately 0.1 mm diameter), colourless and negatively buoyant eggs through gonopores located serially along the arms. At Heron Reef, *L. laevigata* showed a positive response to treatment (approximately 1 million eggs shed) from mid-winter to mid-summer. Spontaneous spawning occurred only in the summer months (November, December).

Females of *Nardoa novaecaledoniae* release approximately 1000, large (approximately 1.0 mm diameter), orange and positively buoyant eggs from gonopores arranged serially in each arm. The eggs undergo lecithotrophic development. The peak of sexual activity occurred in late summer (January) but specimens dissected in mid-winter (July) showed extensive gonad development. No response to 1-methyl adenine injection could be produced at this time. It would appear that this species can spawn and undergo complete gametogenesis within six months, but spawning was observed only once a year.

Females of *Nardoa pauciforis* produce eggs which appear very similar to that of *Nardoa novaecaledoniae*. They are the two common species of this genus on the Great Barrier Reef and are both very similar as adults. They are distinguished by the compression of the distal plates of the arms in *Nardoa novaecaledoniae*. The peak of sexual activity occurred one month earlier (December) in *Nardoa pauciforis* than in *Nardoa novaecaledoniae*, indicating a degree of reproductive isolation in these species at the southern end of the Great Barrier Reef.

Ophidiaster granifer produces eggs which undergo parthenogenetic development (Yamaguchi and Lucas, 1984) and in this study only females (total of 7 individuals) were observed to spawn. Small numbers (20-60 per female) of large (0.6 mm diameter), neutrally buoyant, bright red eggs underwent at least initial development despite no obvious sperm having been released in the water. The only spawning activity in this species was observed in early summer (November, December).

Disasterina abnormalis does not appear to be abundant on the Great Barrier Reef other than in the Capricorn Group at its southern end. At Heron Island this species is abundant behind an extensive rubble bank on the northern side of the reef. The eggs of this species are small (approximately 0.1 mm diameter), colourless and sticky. They sank to the bottom of the aquarium and adhered to the glass, from which they were hard to dislodge. The type of development is unknown. The peak of sexual activity occurred in late spring (October).

Echinaster luzonicus liberates small numbers (20-100 per female) of positively buoyant and approximately 1.0 mm in diameter red eggs during late summer (February). Specimens with fully developed arms were selected for the injection of hormone as arm regeneration following autotomy, which is common in this species, may be at the expense of gonad development.

The remaining species either showed no response to 1-methyl adenine or were not sampled in sufficient numbers to establish reproductive periodicity. The following responses to treatment occurred:

Iconaster longimanus	no response July (N=1)
Culcita novaeguineae	one female February (N=4) no response October (N=4)
Asteropsis carinifera	no response July (N=1)
Gomophia egyptiaca	two males and one female December (N=3) no response February (N=1), May (N=1), July (N=2)
Linckia multifora	no response throughout study (N=120)
Nardoa rosea	one male February (N=1) no response July (N=2), November (N=4)
Ophidiaster armatus	one male July (N=2) no response February (N=1), October (N=1), December (N=1)
Ophidiaster confertus	no response July (N=1)
Ophidiaster robillardi	no response May (N=7), June (N=1), October (N=4), November (N=1), December (N=3)
Tamaria megaloplax	no response July (N=1)
Asterina burtoni	no response throughout study (N=59)
Echinaster stereosomus	no response July (N=2)

Table 6.9 The primary type of reproduction (REPRO) and the type of larval development (DEVEL) where known are shown for all asteroid species recorded from Heron Reef.
- = not known; PLANK = Planktotrophic; LECITH = Lecithotrophic
A = Achituv (1972); B = Barker (1977); Y = Yamaguchi (1975);
* = this study

SPECIES	REPRO.	DEVEL.	SOURCE
Astropecten polyacanthus	SEXUAL	PLANK	Y
Iconaster longimanus	-	-	
Culcita novaeguineae	SEXUAL	PLANK	Y
Acanthaster planci	SEXUAL	PLANK	Y
Asteropsis carinifera	SEXUAL	PLANK	Y
Dactylosaster cylindricus	-	-	
Fromia elegans	SEXUAL	LECITH	*
Fromia milleporella	-	-	
Gomophia egyptiaca	SEXUAL	LECITH	Y
Linckia guildingii	ASEXUAL	PLANK	*
Linckia laevigata	SEXUAL	PLANK	Y
Linckia multifora	ASEXUAL	PLANK	Y
Nardoa novaecaledoniae	SEXUAL	LECITH	*
Nardoa pauciforis	SEXUAL	LECITH	*
Nardoa rosea	SEXUAL	LECITH	*
Neoferdina cumingi	-	-	
Ophidiaster armatus	-	-	
Ophidiaster confertus	-	-	
Ophidiaster granifer	SEXUAL	LECITH	Y
Ophidiaster lioderma	-	-	
Ophidiaster robillardi	ASEXUAL	-	
Ophidiaster watsoni	-	-	
Anseropoda rosacea	-	-	
Asterina anomala	ASEXUAL	-	A
Asterina burtoni	SEXUAL	-	A
Disasterina abnormalis	SEXUAL	-	*
Disasterina leptalacantha	-	-	
Tegulaster emburyi	-	-	
Mithrodia clavigera	SEXUAL	PLANK	Y
Echinaster luzonicus	ASEXUAL	LECITH	*
Coscinasterias calamaria	ASEXUAL	PLANK	B

6.4 Discussion

The species of coral-reef asteroids studied at Heron Island showed differences in the length of their breeding season and this may reflect on their colonisation ability (Mileikovsky, 1971). The length of the breeding season within a species might vary with latitude and the further the population is from the equator, the shorter may be the season for summer breeders. However, except for *Linckia laevigata*, the species studied at Heron Reef generally showed a one to two month breeding season. In two species, *Linckia multifora* and *Asterina burtoni*, no sexual activity was observed throughout the study. This might result from lower than required water temperature at Heron Reef for most of the year (see Mladenov *et al.*, 1986). In this regard, Mortensen (1937) was able to obtain eggs from *Linckia multifora* in the Red Sea where the water temperature is higher than at Heron Reef. It might also be correlated with an increased emphasis on asexual reproduction in *Linckia multifora* once a reef has been colonised by a few sexually reproduced individuals. Although *Asterina burtoni* was not observed to undergo either sexual or asexual reproduction, the distinction between *A. burtoni* and the small fissiparous *A. anomala* is unclear. It is possible that *A. anomala* is an asexually reproducing form of *A. burtoni*.

Nardoa novaecaledoniae and *N. pauciforis* possess arms swollen with gametes for much of the year but still have only a limited breeding season. At Heron Reef, at the southern end of the Great Barrier Reef, *Nardoa pauciforis* is reproductively mature earlier in the summer than is *Nardoa novaecaledoniae* and its eggs are released in November or early December. At this time *Nardoa novaecaledoniae* is not capable of releasing eggs and sperm. Although both species look similar they appear to have limited interbreeding, at least over part of their geographic range. In general, temperature seems to be an important factor in gametogenesis, but the final spawning trigger is dependent on lunar/tidal cycles in many species

(Pearse, 1970, 1975; Yamaguchi and Lucas, 1984).

In coral-reef asteroids the range in fecundity is extremely large. *Fromia elegans*, *Gomophia egyptiaca*, all species of *Nardoa*, *Ophidiaster granifer* and *Echinaster luzonicus* produce large, buoyant, highly pigmented and yolky eggs. While larval development was not studied in detail, the initial phases of lecithotrophic development were observed in these species. The large reserves of yolk should ensure that the resulting larvae need not feed while in the plankton. The number of eggs produced with this development was not studied in detail but appeared to be relatively small (that is, less than 1000 and sometimes much fewer per individual). The eggs are buoyant, opaque and 0.6 to 2.0 mm in diameter. For any species, the egg size, in combination with number of eggs liberated, is an index of reproductive effort. The energetic fecundity in relation to body size of different species, might represent qualitatively different reproductive strategies.

Planktotrophic larvae are produced by *Astropecten polyacanthus*, *Choriaster granulatus*, *Culcita novaeguineae*, *Acanthaster planci*, *Asteropsis carinifera*, all species of *Linckia*, *Mithrodia clavigera*, *Leiaster leachi* and *Coscinasterias calamaria* (Yamaguchi, 1975; Barker, 1977). With this type of larval development, many (up to 1 million), small (0.1 to 0.2 mm), non-yolky, transparent eggs are produced. These eggs appeared less buoyant than eggs that contain large yolk reserves. This may influence dispersion.

Larvae of species of *Nardoa* and other genera which undergo lecithotrophic development may be less likely to die of starvation in the plankton compared with those of species that undergo planktotrophic development and have an obligate larval feeding stage before settlement (see e.g. Thorson, 1950, 1966; Vance, 1973; Barker, 1977; Strathmann and Vedder, 1977; McEdwards and Janies, 1993). Lecithotrophic larvae have shorter development times, but the planktonic stage can be extended if suitable settlement sites are not available

(Strathmann, 1978; Yamaguchi, 1974; Yamaguchi and Lucas, 1984). However, these larvae cannot remain in the plankton for as long as larvae with planktotrophic development. Their dispersal ability and genetic exchange is lower (Scheltema, 1968, 1971; Nishida and Lucas, 1988; Nash et al., 1988; Mladenov and Emson, 1990; Benzie and Stoddart, 1992 a,b). Yamaguchi (1975 b) has commented on the low abundance of lecithotrophic species on oceanic atolls. However, such species are well represented at Heron Reef, a situation that might result from the proximity of adjacent reefs, which would allow short lived larvae from one reef to settle on nearby reefs (see Fisk and Harriott, 1990). Any species may have difficulty colonising over distances greater than its larval dispersal capacity and lecithotrophic species might suffer local extinction following large scale destruction or alteration of coral reef habitat.

Despite the high sexual reproductive effort displayed by most of the large-bodied species, there is little evidence of high recruitment of starfish at Heron Reef. Loosanoff (1964) observed periodic high recruitment during a 25 year study of the temperate species *Asterias forbesi*. Periodic high recruitment has also been observed in the coral reef species *Acanthaster planci*.

It is possible that many eggs are never fertilised when adult populations exist at low densities, such as at Heron Reef. Many fertilised eggs or subsequent larvae would also die from predation or starvation in the plankton (Jackson and Strathmann, 1981; Olsen, 1987). The availability of suitable settling substrate or post-settlement benthic predation might also limit the recruitment of juveniles.

It might be expected that planktonic regulation would be less constant than benthic regulation because of the relative unpredictability of small scale water circulation and the extremely patchy distribution of planktonic predators compared with a possibly more regular and species-specific mortality caused by benthic predators. The post-settlement survival of small juvenile *Acanthaster planci* was examined by Keesing and Halford (1992) and Keesing and Cartwright (1993) who found a difference in survivorship between caged specimens compared with uncaged specimens. For the less common species at Heron Reef, it is possible that many of their eggs are not fertilised. Adult numbers will be further regulated by a combination of either larval starvation or larval and juvenile mortality.

When the energy content of a liberated egg is considered, two different reproductive strategies are apparent. Sexual recruitment can follow either planktotrophic or lecithotrophic larval development (Hendler, 1975; Yamaguchi, 1977 b; Lessios, 1990; McEdwards and Chia, 1991). Because many small eggs can represent the same investment of energy as a few large eggs, the energetic fecundity per unit body weight can be similar in both strategies, despite the difference in numerical fecundity.

It has been suggested (Vance, 1973; Yamaguchi, 1973 a, 1973 b, 1977 b) that lecithotrophic development is an adaptation to high predation or starvation of larva. With this development it is possible the length of larval life can be shorter and hence larval survival should be favoured. On oceanic atolls, species with lecithotrophic development are never abundant and this could result from their poor dispersal ability (Yamaguchi, 1975 b). On Heron Reef, and possibly the Great Barrier Reef in general, where many reefs exist in relatively close proximity, lecithotrophic genera such as *Nardoa*, *Fromia* and *Echinaster* appear to be more abundant than they are on atolls. However, planktotrophic development is favoured where high dispersal is required or when planktonic predation is low and planktonic food is predictable (Menge, 1975; Mileikovsky, 1971; Vance, 1973; Yamaguchi, 1977 b). Many species with this type of larval development occur on oceanic atolls but they occur also on the reefs of the Great Barrier Reef.

Disasterina abnormalis liberated small sticky eggs which sank and adhered to the substrate. The type of development was not studied, but the low dispersion capacity of its eggs might explain its apparently limited distribution. This species was abundant locally. In all other sexually reproducing species, juveniles were uncommon and there was no evidence of either periodic high recruitment or mortality.

The extent of larval dispersion is an important factor in our understanding of the community dynamics within a reef or reef system. If high between-reef larval dispersal occurs, then the adaptive significance of the dispersal phase is the location of spatially and temporally isolated patches in the survival mosaic of each species. Alternately, if larvae recruit primarily into the parent reef population, as a result of circular water movement patterns (Atkinson, Smith and Stroup, 1982; Dight *et al.*, 1990 a, b; Black and Moran, 1991; Black, 1993 but see Wolanski, 1993), then the adaptive significance of the dispersal phase is the avoidance of planktonic or benthic predation in shallow water.

The development time of one month, possessed by many asteroid larvae with planktotrophic development (Yamaguchi, 1977 b; Williams and Benzie, 1993), allows potentially high dispersal, and this development time can be extended further if no suitable settlement site is available. Larvae with lecithotrophic development are capable also of extending the length of the pelagic phase (Yamaguchi, 1974; Yamaguchi and Lucas, 1984). The larvae of coral-reef starfish generally require a solid substrate to complete their development, and a coralline algal substrate has been observed as the chosen settling surface for many species (Yamaguchi, 1973 b; Johnson *et al.*, 1991). More complex species specific signals, located by sensitive chemosensory receptors might ensure settlement in habitats which are conducive to survival of post-settlement stages (Burkenroad, 1957; Morse, 1984). Yamaguchi (1977 c) showed that some juvenile starfish have exponential growth during the period following settlement and proposed that juveniles are subject to high mortality during this period. The juveniles transform to adult morphology at a certain size and before this size is attained may look quite different from adults (e.g. *Culcita novaeguineae* illustrated by Clark, 1921).

The phenomenon of aggregation (Ormond *et al.*, 1973), parthenogenetic development (Yamaguchi and Lucas, 1984), hermaphroditism (Achituv, 1972) or asexual reproduction (Rideout, 1978) may be correlated with survival at low population densities. In low density, spatially dispersed populations of starfish, there is a low probability of locating a conspecific of the opposite sex at breeding time.

CHAPTER 7

ASEXUAL REPRODUCTION

7.1 Introduction

Some species of coral-reef Asteroidea are known to exhibit both sexual and asexual reproduction (Yamaguchi, 1975 b). The species known to reproduce asexually are detailed by Emson and Wilkie (1980). Rideout (1978) has shown that asexual reproduction is the chief form of reproduction in the asteroid *Linckia multifora* at Guam. Achituv and Sher (1991) have suggested that *Asterina burtoni* reproduces only by asexual reproduction in the Mediterranean Sea. However, the relative roles of sexual and asexual reproduction in the population maintenance of other coral-reef asteroid species have not been studied. It is not known whether asexually reproducing species have a regular alternation of sexual and asexual activity, or if sexual reproduction is strongly reduced or even absent in any of these species.

In asteroids, asexual reproduction involves either the splitting of the disc (fission), or the casting off of arms that regenerate new starfish (autotomy). In some species, autotomised arms need not contain any section of disc or madreporite for their survival (Clark, 1913; Edmondson, 1935). In such species, regeneration of a mouth and basic digestive organs, must occur while the regenerating arm is metabolising stored reserves of energy (Lawrence *et al.*, 1986). A description of the stages of regeneration, following autotomy in *Linckia multifora*, is provided by Rideout (1978).

The period of regeneration following autotomy, before the regenerated arms are ready for reautotomy, might be prolonged greatly in species larger than *Linckia multifora*. Autotomy is reduced when individuals are infected with the parasitic gastropod *Stylifer* as this parasite inhibits autotomy of infected arms. Because the mortality of regenerating individuals is high (Davis, 1967; Rideout, 1978), though not as great as expected by Clark (1913), this inhibition of autotomy results in greater survival of the parasite (Davis, 1967).

An alternation of sexual and asexual activity has been observed in *Nepanthia belcheri* (Ottesen and Lucas, 1982). Some sexual activity has been recorded in the asexually reproducing species, *Ophidiaster robillardi* (Yamaguchi and Lucas, 1984). Other species, such as *Asterina anomala* and *Linckia multifora*, often remain small and sexually immature through the continuing process of asexual autotomy or fission. Some sexual activity has been noticed in *Linckia multifora* and *Linckia guildingii* (Mortensen, 1937, 1938) but the contribution of this as a means of population maintenance may be outweighed by that of autotomy (Rideout, 1978).

7.2 Methods

Individuals of any species that exhibited signs of recent autotomy or fission were identified, collected and measured. The presence of comets was considered indicative of asexual reproduction by means of autotomy, and recent disc fission, followed by regeneration of more than half the disc, indicative of reproductive binary fission. The frequency of occurrence of asexual products was recorded for each sampling period. Measurements were taken routinely of the number of arms and the length of the longest arm of each individual. Periodically, all arms of asexual specimens were measured and details of obvious regeneration or autotomy were recorded.

In *Linckia multifora* and *Echinaster luzonicus*, the mean major radius (mean R mm), mean minor radius (mean r mm) and mean major/minor radius (R/r) were calculated for each sampling period. The variation throughout the study in mean major radius of these species is discussed further in Chapter 8.

7.3 Results

Six species of asteroid occurring at Heron Reef exhibited signs of asexual reproduction either by autotomy or binary fission. Specimens of *Linckia guildingii*, *Linckia multifora*, *Echinaster luzonicus* and *Ophidiaster robillardi* were observed in varying stages of regeneration following arm autotomy. Specimens of *Asterina anomala* and *Coscinasterias calamaria* were observed in varying stages of regeneration following binary fission. Autotomous species can regrow a complete individual from the distal half of an arm, with no need for any portion of the disc or madreporite to be included.

The first stage of regeneration in the autotomised arm is called a comet after its characteristic shape. Comets were encountered frequently in these species and most of the individuals collected in this study had recently autotomised at least one arm, with the remaining arms being in various stages of regeneration. Individuals with all arms of equal length were very uncommon. On several occasions, recently autotomised arms were found alongside the parent animal in the field. In large specimens of *Echinaster luzonicus*, the process of autotomy can proceed very quickly, and if the animals are handled roughly, arms can be autotomised within seconds. If high water quality and a temperature comparable with that which occurs naturally, are not maintained when *Linckia multifora* and *Echinaster luzonicus* are kept in aquaria, freshly autotomised arms die.

During the period of study, large changes in mean major radius were observed in both *Linckia multifora* and *Echinaster luzonicus*. At times of smaller mean individual size, the frequency of comet stages in the population was generally higher, but this varied between years. In 1978, the number of comet *Echinaster luzonicus* found in the field corresponded well with periods of lower mean individual size. For example, in May 1978, 29% of individuals were comets, whereas in August

1978, the proportion of comets was only 15%. This declined to about 5% in June 1979, and did not vary greatly over the remainder of the study. In *Linckia multifora* the proportion of comets in May, August and November 1978 was 32%. This declined to 8% over the following two years. In 1981 and 1982, the proportion of comets increased to 18%.

The difficulty in locating comet stages will have biased these results. Because the frequency of comets and autotomised limbs was difficult to both sample and analyse (and was not studied in detail), variation in the mean major radius (R) and mean major/minor radius (R/r) was considered a more reliable index of the frequency of autotomy.

In both *Linckia multifora* and *Echinaster luzonicus*, variation in the mean major radius (R) and mean major / minor radius (R/r) is illustrated in Figures 7.1 and 7.2. The significance of this variation in mean major radius (R) is discussed further in Chapter 8. While specimens of *Linckia guildingii* and *Ophidiaster robillardi* that had recently undergone autotomy were observed throughout the study, only *Linckia multifora* and *Echinaster luzonicus* were common enough for this variation in mean size to be analysed.

At Heron Reef, *Asterina anomala* and *Coscinasterias calamaria* reproduce by binary fission which results in two halves each regrowing to form two complete individuals. In this case a portion of disc containing a madreporite is always present as both species possess several madreporites on the disc.

The remaining species in the Heron Reef asteroid assemblage, while possessing great powers of regeneration, showed no evidence of using autotomy or fission as a form of asexual reproduction. If parts of the body of these species are autotomised, these parts die and the remaining body regenerates the lost limbs.

Figure 7.1 *Linckia multifora*
Variation in both mean major radius (R mm) and mean major / minor radius (R/r) during study period.

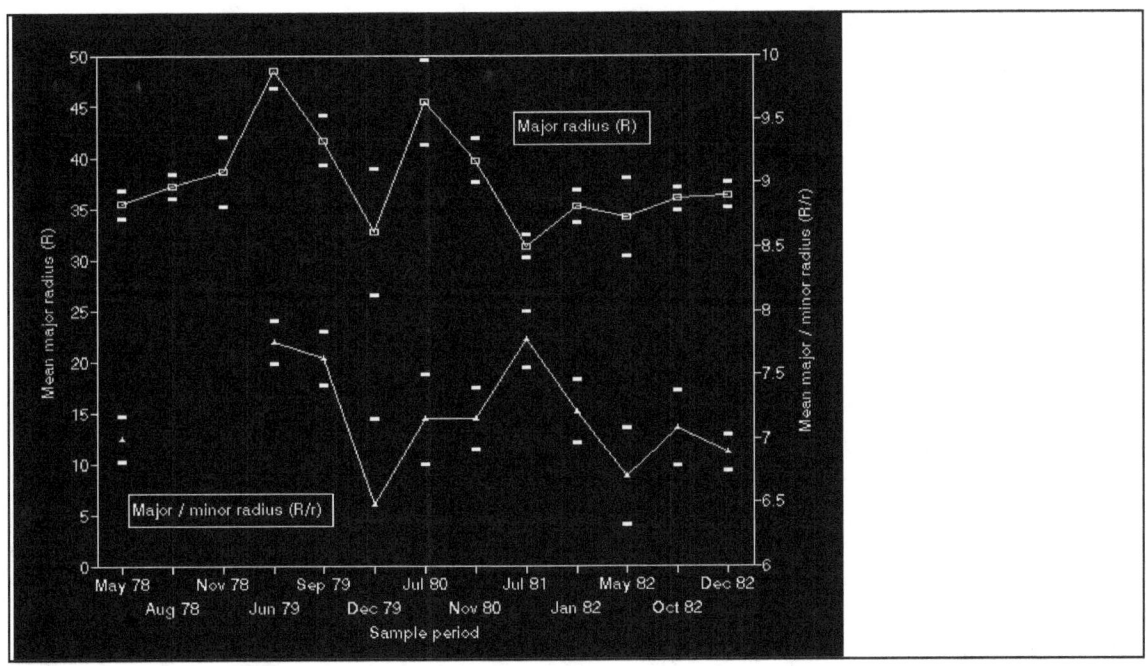

Figure 7.2 *Echinaster luzonicus*
Variation in both mean major radius (R mm) and mean major / minor radius (R/r) during study period.

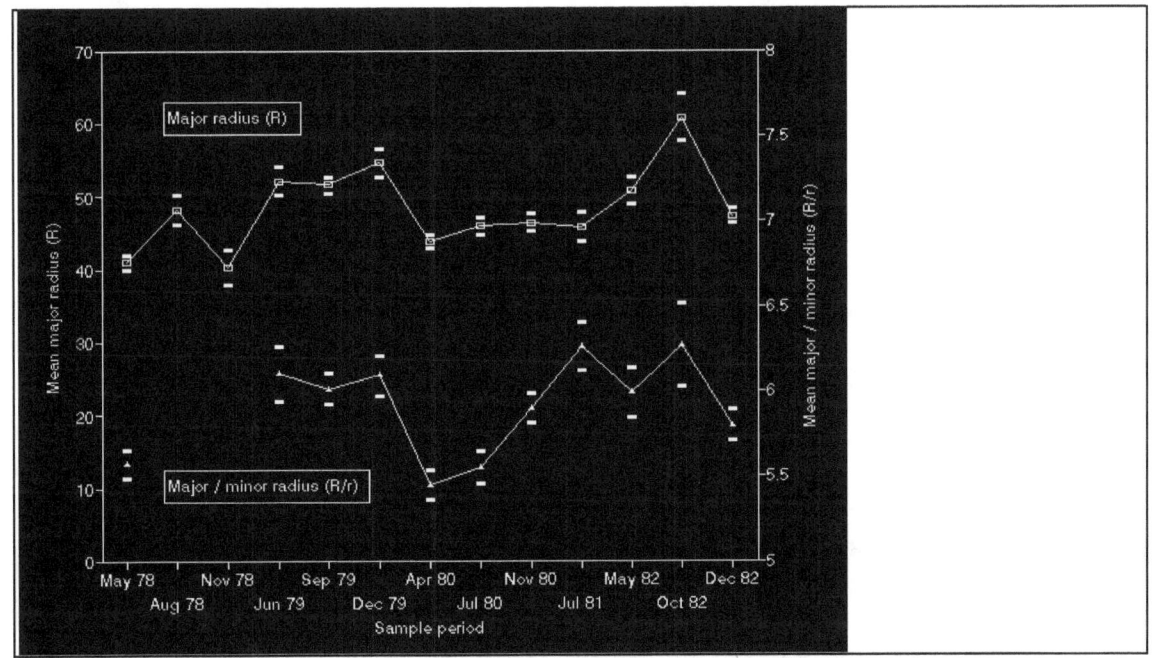

7.4 Discussion

Asexual reproduction can offset the effects of intense benthic and planktonic larval predation, as well as those caused by the general vicissitudes of planktonic life which include dispersal loss and starvation (see Yamaguchi, 1975 b; Rideout, 1978; Franklin, 1980; Ottesen and Lucas, 1982; Yamaguchi and Lucas, 1984; Olsen, 1987). On coral reefs, benthic predation of larvae and newly settled recruits might be too high for benthic larval development or brooding behaviour (see Menge, 1975; McClary and Mladenov, 1989; McClary, 1990; Bosch, 1989; Bosch and Pearse, 1990; Komatsu et al., 1990) to be a viable survival strategy.

Two modes of reproduction are employed by some asteroids, one mode allowing the species to disperse, the other, the build-up of population numbers once colonisation is established (Yamaguchi and Lucas, 1984). Cameron and Endean (1982) suggested that autotomy is an adaptation to predation and Birkeland et al (1982) observed autotomy in their study of asteroid predatory interactions. A number of tropical and temperate asteroids are known to undergo regular autotomy (Yamaguchi, 1975 b; Rideout, 1978; Emson and Wilkie, 1980; Crump and Barker, 1985; Mladenov et al., 1989; Dubois and Jangoux, 1990).

The size distribution of *Linckia multifora* and *Echinaster luzonicus* varied significantly over the study period (see Table 8.1). While the autotomy rate varied within and between years, the frequency of comet stages in the population was also determined by the survival rate of autotomised arms. This seemed to vary considerably from one year to another. Although comet stages of *Linckia guildingii* and *Ophidiaster robillardi* were found, there was no obvious temporal variation in their occurrence. The abundance of these species was not sufficient to allow analysis of the change in mean individual size. Although individuals of *Asterina anomala* and *Coscinasterias calamaria* were found in varying stages of regeneration, once again, there appeared to be no temporal pattern in the occurrence of asexual stages (compare with Muenchow, 1978; Ottesen and Lucas, 1982). All specimens of *Asterina anomala* were small and fissiparous and this species appeared to be distinct from *Asterina burtoni*, at Heron Reef.

At Heron Reef, sexual reproductive effort appears to be very low in *Echinaster luzonicus* and *Linckia guildingii*. No sexual activity was recorded in *Linckia multifora* at all. Small individuals of these three species resulted invariably from autotomy. It would appear that asexual reproduction is the chief means of population maintenance in these three species. *Ophidiaster robillardi* is less common than either *Linckia guildingii*, *Linckia multifora* or *Echinaster luzonicus* at Heron Reef. On adjacent Wistari Reef, the density of *Ophidiaster robillardi* in small patches on the reef crest was considerably higher than at Heron Reef. The spatial distributions of *Linckia multifora*, *Ophidiaster robillardi* and *Echinaster luzonicus* were highly clumped. This patchiness in abundance might be a result of local population increases following an initial sexual colonisation (Ottesen and Lucas, 1982; Mladenov and Emson, 1984; 1990; Crump and Barker, 1985; Johnson and Threlfall, 1987). *Linckia multifora* and *Echinaster luzonicus*, together with *Disasterina abnormalis* (which liberates sticky eggs) occurred at the highest local densities recorded during this study.

CHAPTER 8

CONSTANCY OF MEAN SIZE

8.1 Introduction

The fluctuations that occur in animal populations have been regarded as a measure of community stability (MacArthur, 1955; Frank, 1968; Den Boer, 1971; Jacobs, 1974; Goodman, 1975; Brown, 1981). Many authors believe that complex high diversity systems are characterised by relative constancy of species composition (e.g. Dunbar, 1960; 1972; Leigh, 1965; Margalef, 1963; 1974). They propose that populations of the component species do not vary to the extent demonstrated in more simple communities. Additionally, the interaction of competitors and predator / prey situations might prohibit the resource monopolisation so characteristic of dominant species in less diverse systems. Other authors (e.g. Connell, 1978; Sale, 1976, 1977; Sale and Douglas, 1984) believe that there is temporal variability in the community structure of the coral-reef organisms they have studied.

A paucity of juveniles characterises the population structures of large bodied, coral-reef starfish (Yamaguchi, 1973 a). It is possible that populations are maintained either by continual low recruitment or occasional high recruitment, each coupled with iteroparity. The juveniles are cryptic and their apparent absence or rarity indicates that reproductive success is either constantly low, sporadic or both. Amongst coral-reef species, population outbreaks have been well documented for only *Acanthaster planci*. However, population changes in *Linckia laevigata* following *Acanthaster* outbreaks have been suggested by Laxton (1974) and *Asterina burtoni* is known to have extended its range into the Mediterranean Sea following its introduction through the Suez canal (Achituv, 1969).

In most common species of coral-reef asteroid that have been studied, their reproductive strategy was directed towards the production of enormous quantities of gametes which were released directly into the surrounding water (Yamaguchi, 1973 a; 1977 a). If the mortality of the resulting larvae varied greatly from year to year, then we would expect years of noticeable recruitment followed by one or more years of little or no recruitment success. If the adult population is short-lived (e.g. two years), then the recruitment necessary to maintain the population must occur within this short period. If there were another consecutive year of recruitment failure then the species would become locally extinct. In these short-lived species, juveniles should occur in sufficient numbers to be detected. If, however, the adults are long-lived, then the level of recruitment required to maintain the adult population could be extremely low per annum and we might expect to see juveniles only occasionally.

If periods of high recruitment are required to maintain the population structure of a common species, the mean individual size of the populations should vary as a consequence of the influx of juveniles. In *Ophidiaster granifer*, when periods of recruitment occurred at Guam, the mean individual size of the population decreased (see Yamaguchi and Lucas, 1984). The mean individual size should increase progressively throughout the interval between periods of recruitment. In a large bodied species, such as *Linckia laevigata*, the mean individual size should increase slowly (dependent on growth rate), but might reach a size equilibrium determined by the availability of food (Paine, 1976). If periods of recruitment occurred, the mean individual size of the population should decrease. If recruitment did not occur, the mean individual size should increase slowly. The rate of increase in mean individual size will be determined by the average individual growth rate. This might be very slow in a species that is long lived.

8.2 Methods

The size data for the more common species were analysed to see if there was any temporal variation in mean size for each species. A one way ANOVA (ratio of variance in mean major radius (R mm) between and within sampling periods) for each species was calculated and the results are listed in Tables 5.1 - 5.10.

The mean size variation required to produce a probability level of .05 or .01 was very small for these relatively common species. Any probability not less than 0.001 represented only a small mean size variation compared with the size variation within each of the populations.

8.3 Results

Table 8.1
The significance of temporal size variation for each of the more common species. The grand mean (R mm), sample size (N), F statistic, degrees of freedom and probability level of the size variation over several sampling periods are tabled.

SPECIES	MEAN R	N	F	d.f.	P
Linckia guildingii	134	131	1.1	9,82	N/S
Linckia laevigata	127	516	1.7	11,459	<.05
Linckia multifora	38	396	9.5	10,329	<.001
Nardoa novaecaledoniae	88	361	3.2	11,295	<.01
Nardoa pauciforis	104	233	2.2	11,179	<.05
Disasterina abnormalis	15	1109	15.8	12,1054	<.001
Echinaster luzonicus	48	988	11.4	10,883	<.001

Figure 8.1a *Linckia multifora*
Variation in mean major radius (R mm ± S.E.) throughout study

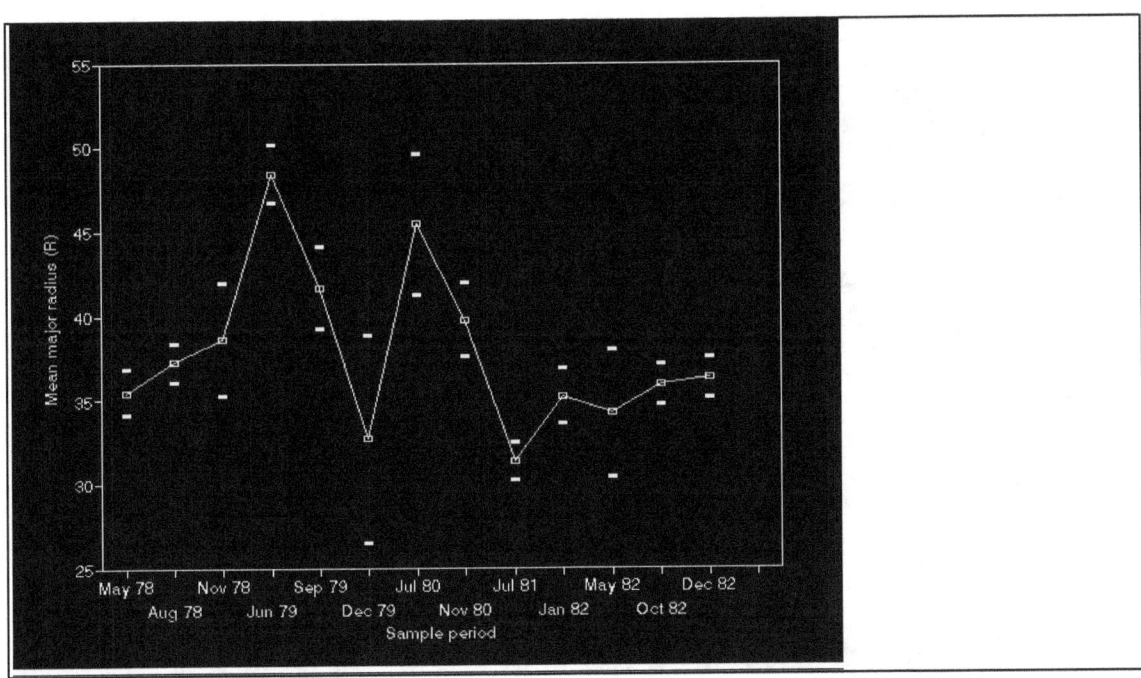

Figure 8.1b *Linckia multifora*
Frequency distribution of major radius (R mm) in May 1978

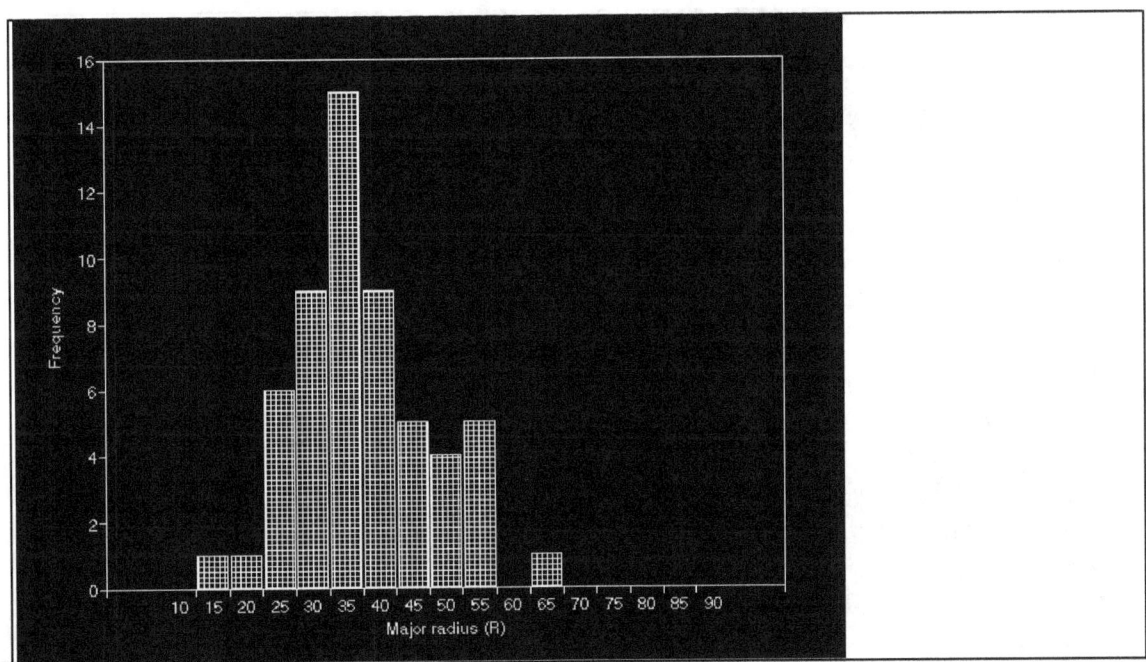

Figure 8.1c *Linckia multifora*
Frequency distribution of major radius (R mm) in June 1979

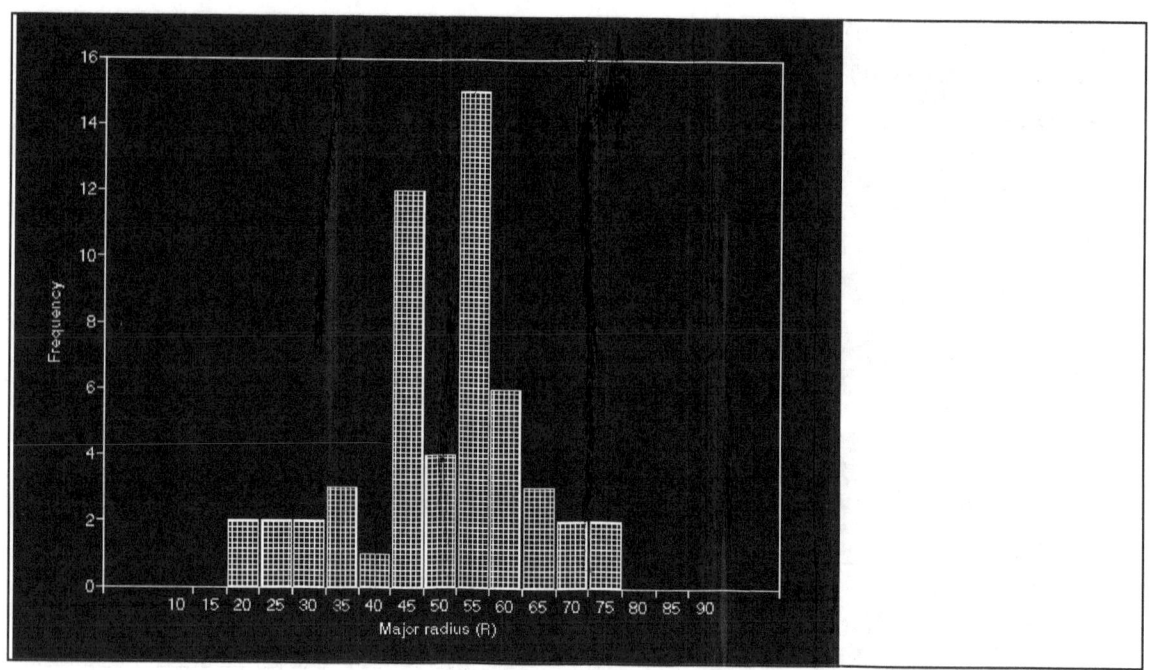

Figure 8.1d *Linckia multifora*
Frequency distribution of major radius (R mm) in July 1981

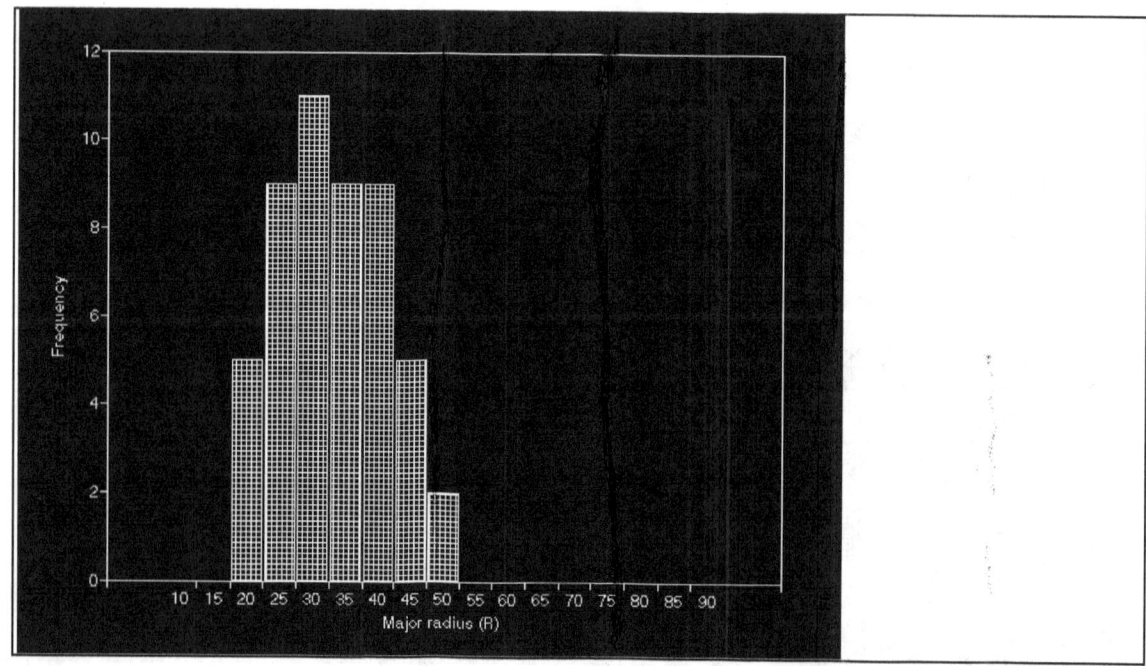

Figure 8.2a *Disasterina abnormalis*
Variation in mean major radius (R mm ± S.E.) throughout study

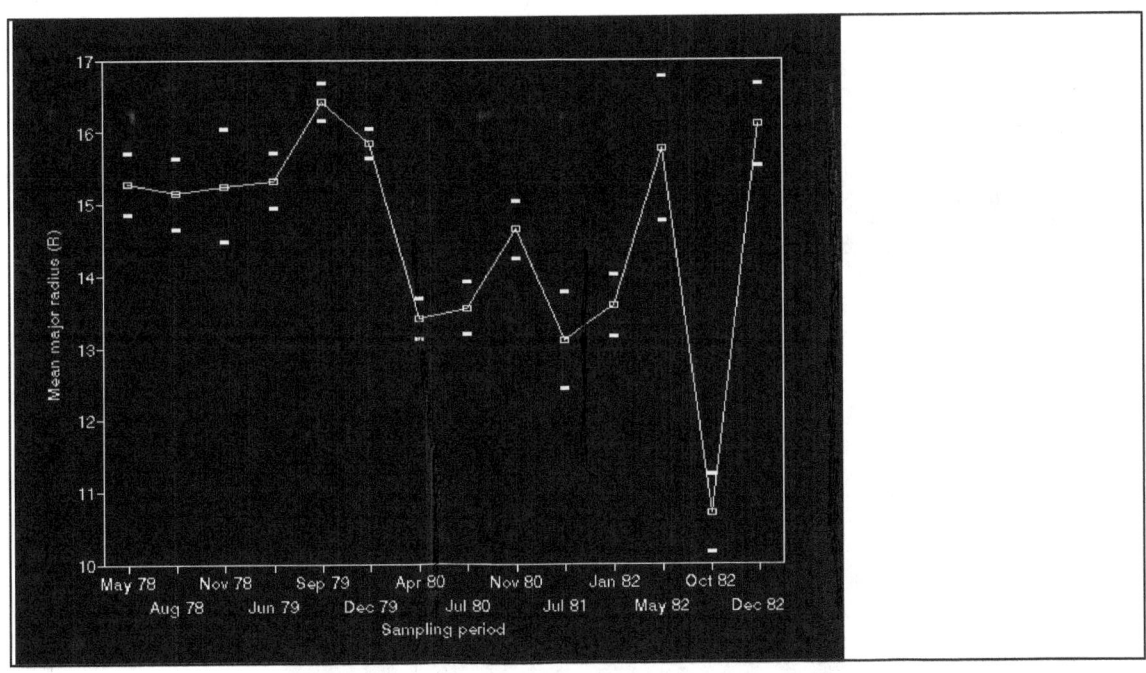

Figure 8.2b *Disasterina abnormalis*
Frequency distribution of major radius (R mm) in May 1978

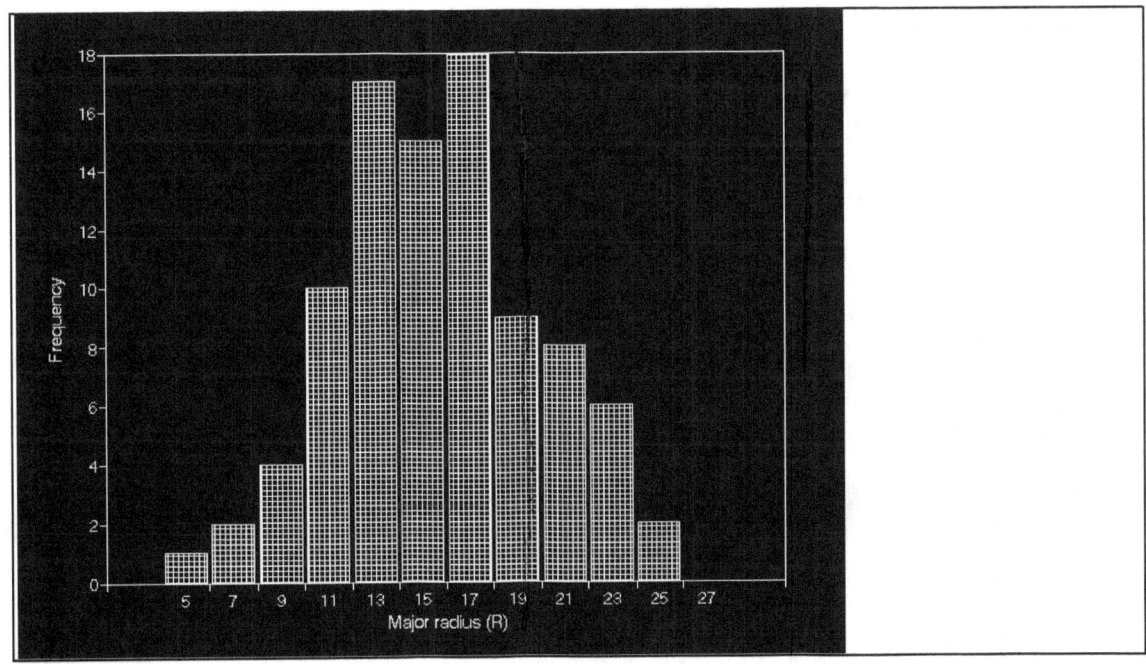

Figure 8.2c *Disasterina abnormalis*
Frequency distribution of major radius (R mm) in September 1979

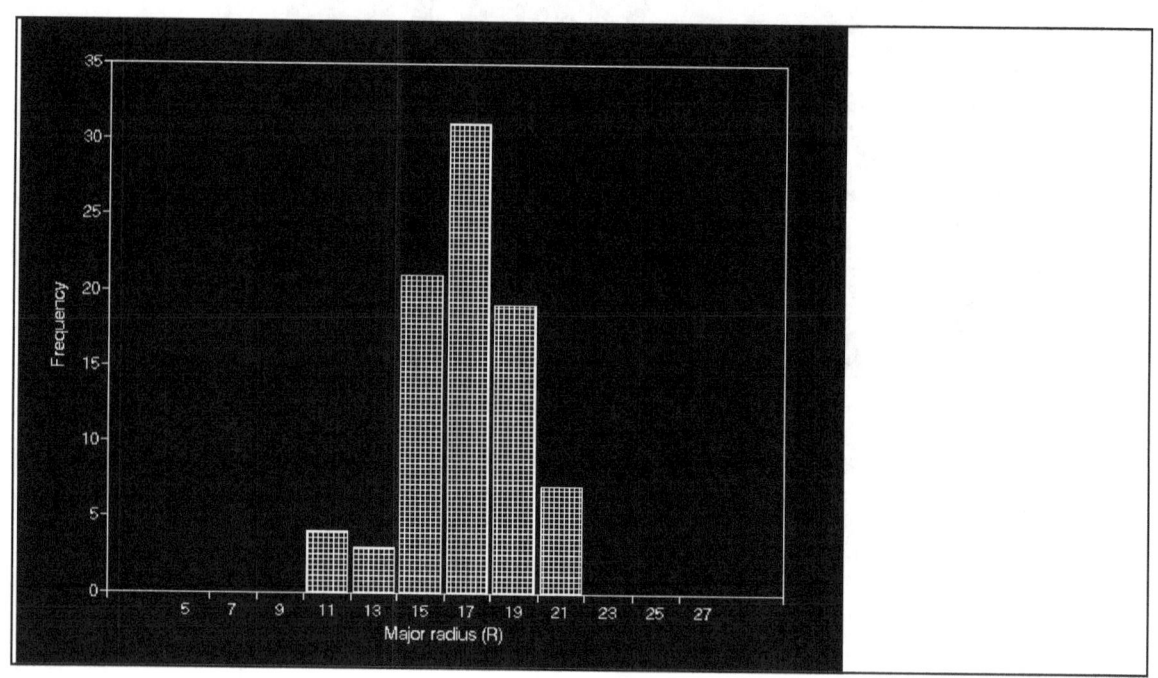

Figure 8.2d *Disasterina abnormalis*
Frequency distribution of major radius (R mm) in April 1980

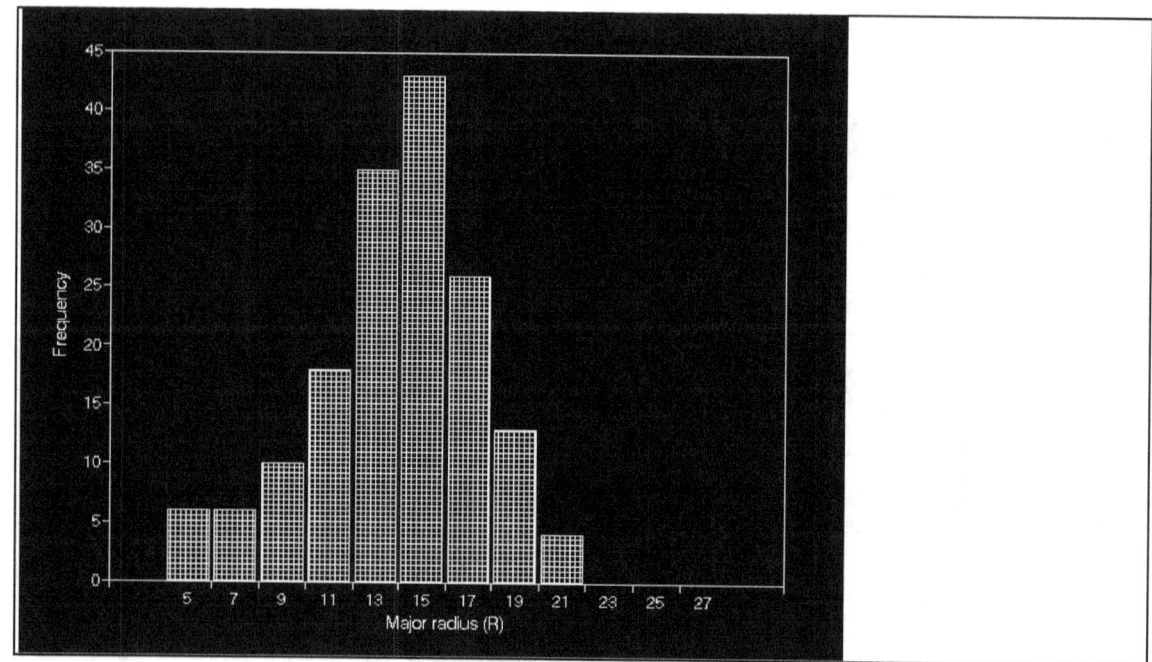

Figure 8.3a *Echinaster luzonicus*
Variation in mean major radius (R mm ± S.E.) throughout study

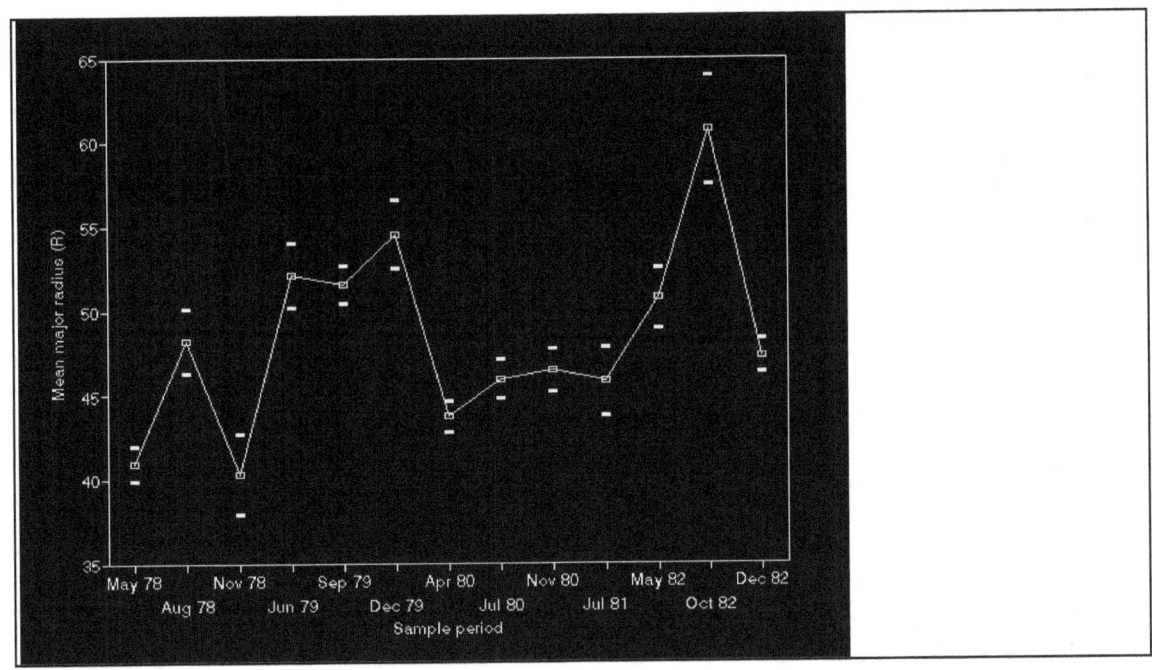

Figure 8.3b *Echinaster luzonicus*
Frequency distribution of major radius (R mm) in May 1978

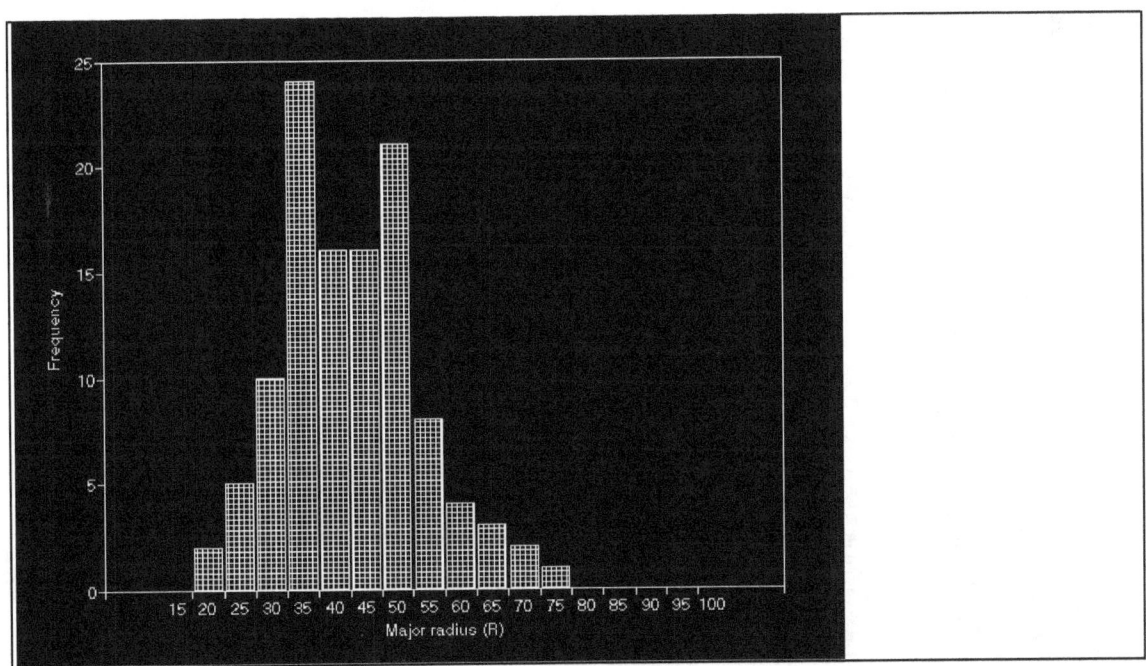

Figure 8.3c *Echinaster luzonicus*
Frequency distribution of major radius (R mm) in August 1978

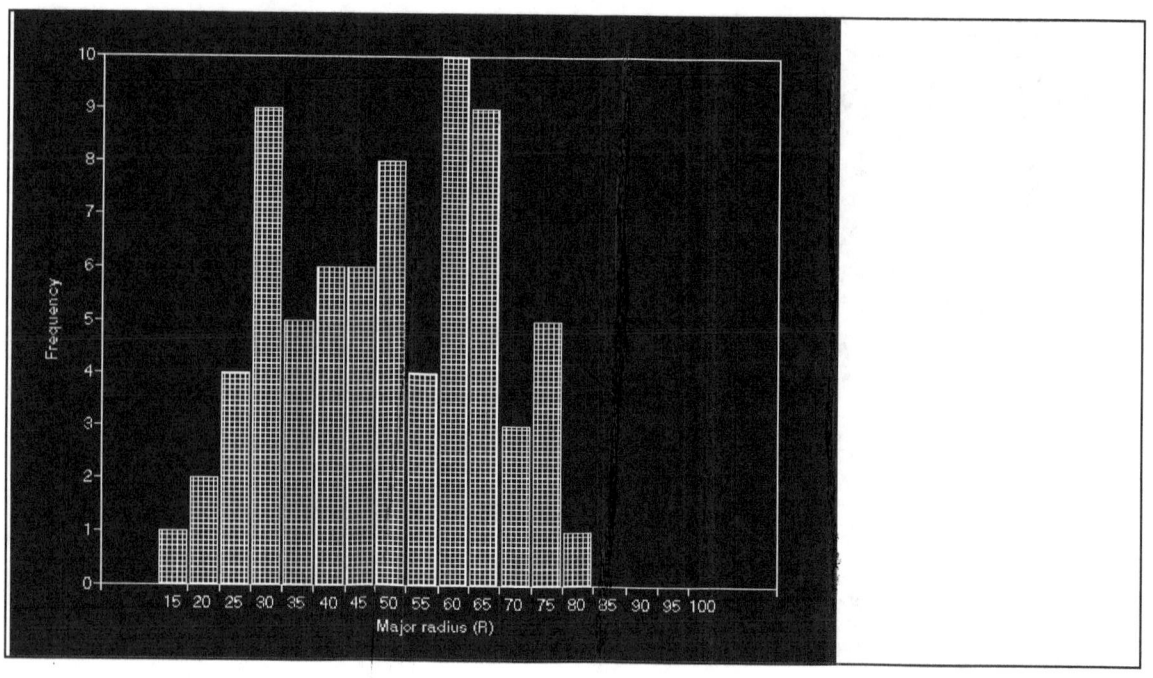

Figure 8.3d *Echinaster luzonicus*
Frequency distribution of major radius (R mm) in December 1979

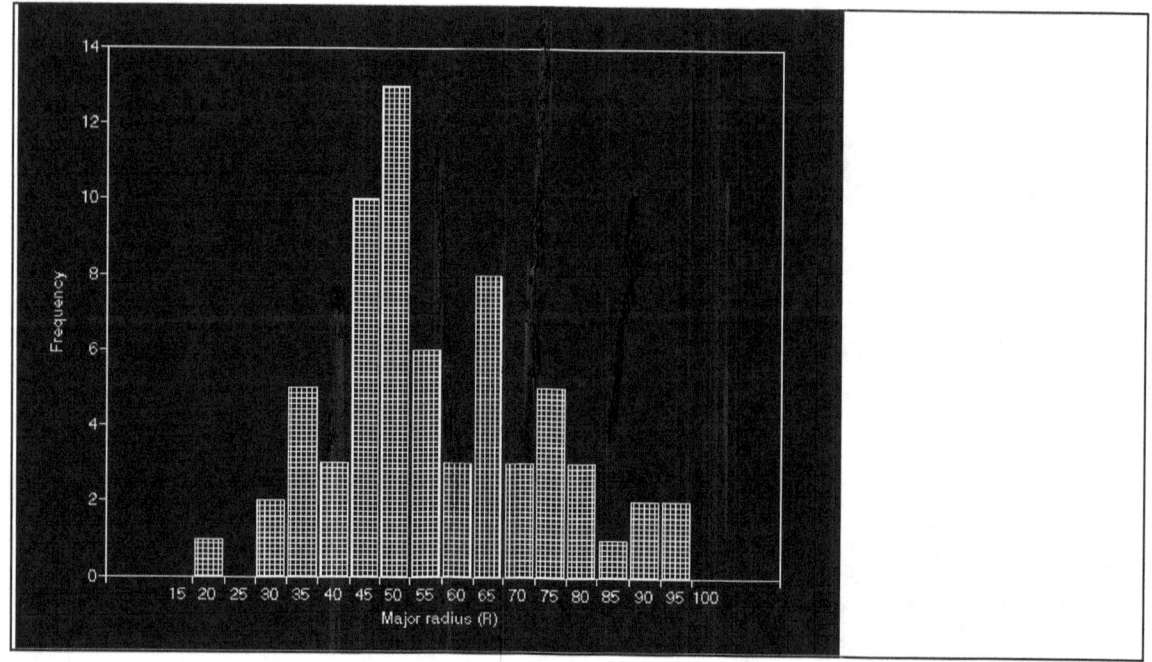

8.4 Discussion

From Tables 5.1 - 5.10, Figures 5.1a - 5.10a and Table 8.1, it can be seen that four of the large-bodied species, *Linckia guildingii*, *Linckia laevigata*, *Nardoa pauciforis* and *Nardoa novaecaledoniae*, did not vary their mean size greatly over the entire study period of five years. In two of these species, *Linckia laevigata* and *Nardoa pauciforis*, the mean size variation was significant at 0.05. In *Nardoa novaecaledoniae*, while significant at 0.01, this variation still represented only a small change in the mean size of the population over the entire study period.

Although four of the seven common species maintained a size distribution that did not vary greatly during the study period, the possibility that many of the species might demonstrate occasional high recruitment success, when observed on a much larger time scale, cannot be rejected. If such recruitment occurred, it would manifest itself as oscillations in the mean individual size of that species. Asteroids are also known to possess highly plastic growth rates which can effectively disguise annual year classes.

It should be noted that a stable size distribution does not necessarily imply low recruitment and low mortality, but can result from a balance of high recruitment and high mortality. Under conditions of high mortality and low recruitment, a population with a low growth rate can also manifest a stable size distribution but it would show a simultaneous decline in population density. This was not observed in the present study and the small change in the mean size of the large-bodied species suggests that *Linckia guildingii*, *Linckia laevigata*, *Nardoa novaecaledoniae* and *Nardoa pauciforis* are long-lived.

During this period, *Linckia multifora*, *Disasterina abnormalis* and *Echinaster luzonicus* showed mean size variations that were highly significant. This size variation was the result of periodicity in either sexual or asexual reproduction. In *Linckia multifora* and *Echinaster luzonicus*, the difference in size resulted from autotomy. High recruitment of juveniles was observed in only one small bodied, sexually reproducing species, *Disasterina abnormalis*. In the remaining species, the abundances were low, and statistically valid comparisons of what might have been temporal mean size variation could not be justified. This applied to *Ophidiaster granifer* that showed periodic recruitment when studied at Guam (Yamaguchi and Lucas, 1984), and *Asterina burtoni* which did not show significant mean size variation in the present study.

The relative stability of the size distributions of the common large-bodied species can be explained by assuming very slow growth of a predominant year class or a balance of recruitment and mortality within each of the species. It seems likely that a combination of both is involved. The paucity of juvenile asteroids, and the constancy of the size distributions in all the large bodied sexually reproducing species can be explained only by a life-history model which incorporates low adult mortality and includes the assumption of longevity.

The variation in mean size between populations of *Linckia laevigata* at different localities in the Indo-West Pacific could be caused by the presence of geographically asynchronous, dominant year classes. However, this is unlikely as this species did not alter its mean size greatly during the period of the present study. The highly plastic growth rate may be influenced by nutrition (see Wolda, 1970) or other factors (e.g. disturbance) may cause both the higher density and smaller mean size. Dwarfism, resulting from high salinity, was described in *Asterina burtoni* by Price (1982).

The results of this study of coral-reef asteroids contrasts with data relating to laboratory rearing of *Acanthaster planci* which are claimed to demonstrate individual senescence at an age of approximately five years (Lucas, 1984). This finding, which appears to be inconsistent with the general biology of an often rare, large-bodied and venomous animal, can be attributable to the laboratory rearing conditions (see Endean and Cameron, 1990 b). Additionally, a specimen of *Acanthaster planci* held in an aquarium at the Heron Island Research Station decreased to two-thirds of its original size within a period of 6 months. When adequate food is not available, regression in size might occur in many coral-reef asteroid species. At Heron Reef, the coral-reef asteroid community is not dominated by violently fluctuating size structures as might be expected from the work of Lucas (1984). All the large-bodied, sexually reproducing asteroids in this study existed with a stable size structure for the entire study period.

CHAPTER 9

RELATIVE ABUNDANCE AND DIVERSITY

9.1 Introduction

In addition to the high diversity of the coral reef ecosystem, a feature of this ecosystem is the large number of rare species within each taxonomic group. The general relation between the number of species and the number of individuals in a sample of a population was discussed by Fisher, Corbet and Williams (1943), who commented that species are not equally abundant, even under conditions of considerable uniformity. They went on to state that the majority of species are comparatively rare while only a few are common.

It is not known whether the rarity of a species is indicative of its low competitive ability or alternately whether the species is restricted to specialised microhabitats with excess recruitment eliminated by predators (Hairston, 1959; Kunin and Gaston, 1993). The relative abundances of the species in a diverse assemblage are often distributed over many orders of magnitude. As a result, qualitative representations of abundance such as common, moderately abundant or rare must be arbitrary in their assignment.

Many different mathematical models have been proposed to describe satisfactorily the relationship that exists between the relative abundances of different species in an assemblage. While each model has been criticised extensively (Hurlbert, 1971; Abbott, 1983; Connor and McCoy, 1979; Connor, McCoy and Cosby, 1983; Martin, 1981; McGuiness, 1984; Pielou, 1981; Sughihara, 1981), each attempts to quantify the degree of variation in the relative abundances of the different species. The most noticeable result of this abundance variation is the different rates at which species accumulate with increased sampling in different assemblages.

9.2 Methods

The population density of each species and the relation between sample area and the number of individuals in the sample was calculated in Chapter 4. The relation between sample area and the total number of species in the sample (the species : area curve) was also calculated from the traverse data. The cumulative number of species was compared with the cumulative area of the traverses (starting at the completion of Traverse 1 and continuing through to the completion of Traverse 72). This comparison was also undertaken with the natural logarithm of the cumulative area of the traverses.

Shannon's Evenness Index (see Pielou, 1981) which is the expression $(\Sigma P(\log P)) / \log S$, where P is the proportion of each species in the community, and S is the total number of different species, is often used to display the relative richness of various communities. Shannon's Evenness was calculated for each traverse individually and cumulatively starting with Traverse 1 and ending with Traverse 72.

The relation between the numerical abundance of each species and the rank abundance of each species was calculated by ordering the numerical abundance from most common (rank 1) to least common (equal rank 20 for five species). Percent relative abundance was the ratio of the numerical abundance of each species to the total asteroid abundance.

9.3 Results

Table 9.1 lists the numerical, relative and rank abundances of each species located on the intertidal traverses. Figure 9.1a graphs the relation between the numerical abundance of a species and its rank abundance. Figure 9.1b graphs the relation between (log) relative abundance and rank abundance. Figures 9.2a,b graph the species : area and species : (log) area relation. Figures 9.3a,b graph the relation between Shannon's Evenness and cumulative area and cumulative (log) area. Natural logarithms were used in all these calculations. Shannon's Evenness as a measure of diversity has the advantage that the index is a ratio of attained diversity over maximum possible diversity and is therefore independent of the base of logarithm which has been chosen.

Table 9.1

The numerical abundance, relative abundance and abundance rank of inter-tidal asteroids at Heron Reef.

SPECIES	NUMERICAL	RELATIVE	RANK
Culcita novaeguineae	15	**	13
Asteropsis carinifera	3	*	19
Dactylosaster cylindricus	1	*	20
Fromia elegans	16	**	12
Fromia milleporella	1	*	20
Gomophia egyptiaca	6	*	16
Linckia guildingii	116	***	8
Linckia laevigata	509	***	3
Linckia multifora	522	***	2
Nardoa novaecaledoniae	326	***	5
Nardoa pauciforis	187	***	7
Nardoa rosea	1	*	20
Ophidiaster armatus	4	*	17
Ophidiaster confertus	4	*	17
Ophidiaster granifer	116	***	9
Ophidiaster lioderma	1	*	20
Ophidiaster robillardi	24	**	10
Asterina anomala	17	**	11
Asterina burtoni	208	***	6
Disasterina abnormalis	500	***	4
Disasterina leptalacantha	7	*	14
Tegulaster emburyi	1	*	20
Echinaster luzonicus	1402	****	1
Coscinasterias calamaria	7	*	14

:*	Very rare	<10	
:**	Rare	11-100	
:***	Common	101-1000	
:****	Abundant	>1000	

Figure 9.1a Relation between number of individuals and rank abundance. Ranks 20 - 24 represent one individual each.

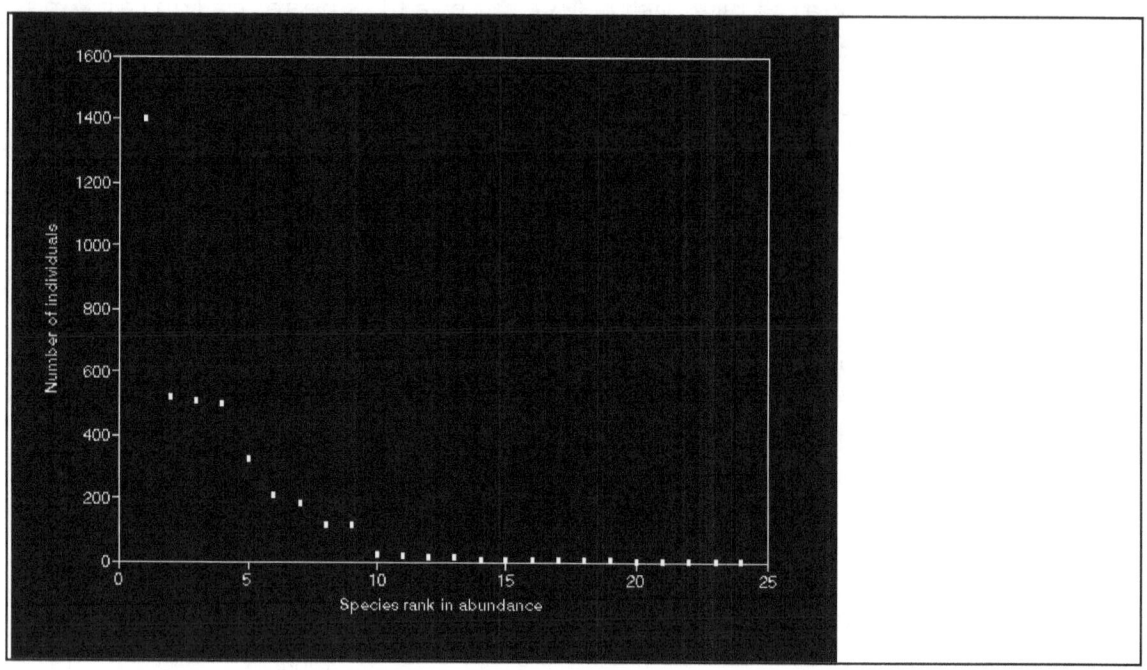

Figure 9.1b Relation between (log) percent relative abundance and rank abundance. Ranks 20 - 24 represent one individual each.

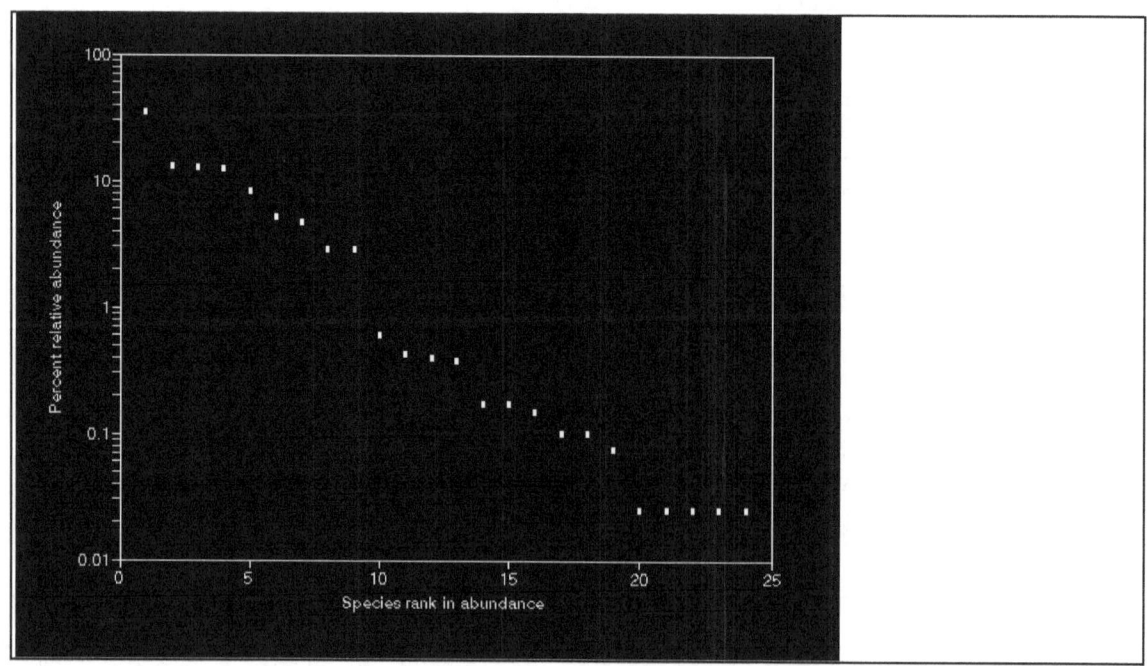

Figure 9.2a Relation between cumulative number of species and area. Traverse 1 to Traverse 72 in chronological order are included.

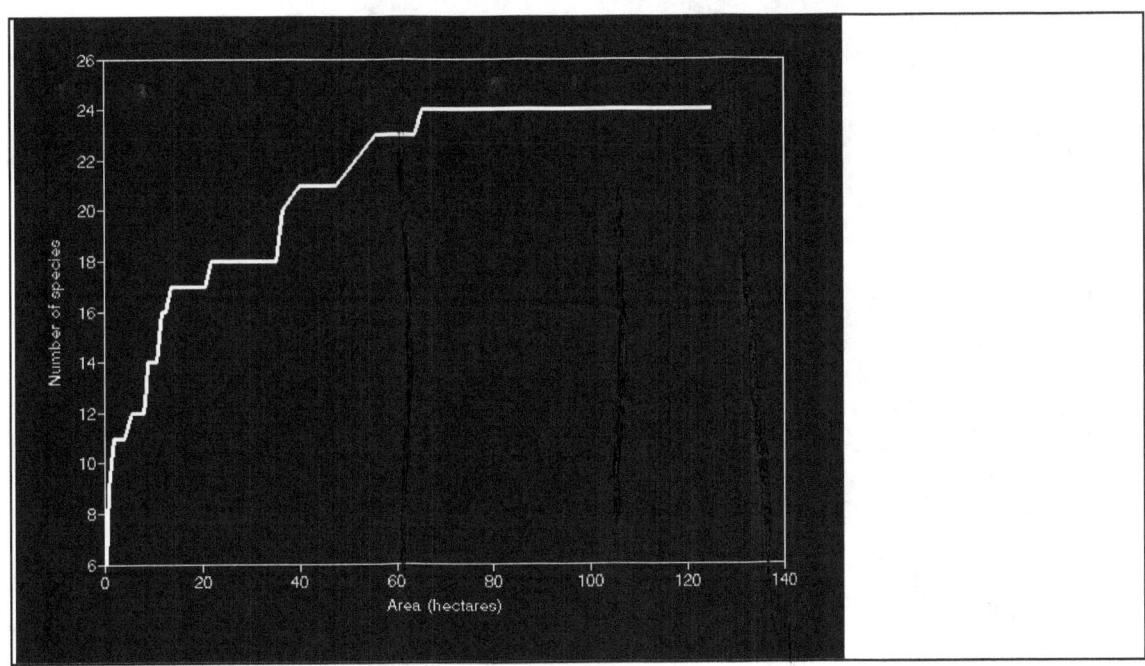

Figure 9.2b Relation between cumulative number of species and (log) area. Traverse 1 to Traverse 72 in chronological order are included.

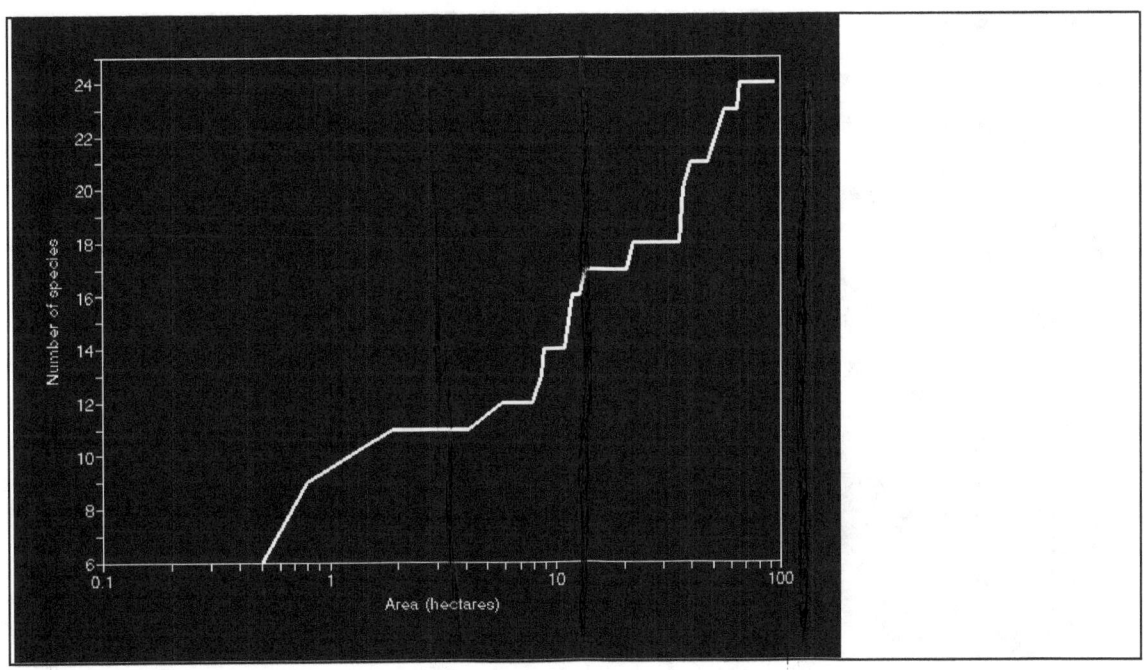

Figure 9.3a Relation between cumulative evenness and area. Traverse 1 to Traverse 72 in chronological order are included.

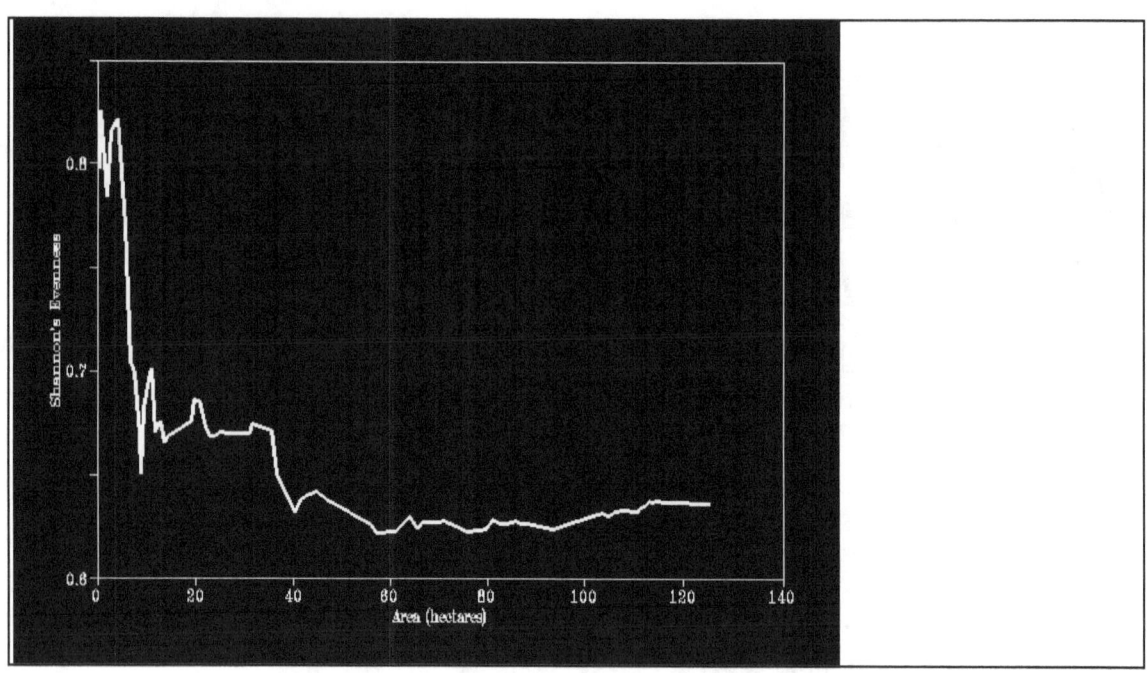

Figure 9.3b Relation between cumulative evenness and (log) area. Traverse 1 to Traverse 72 in chronological order are included.

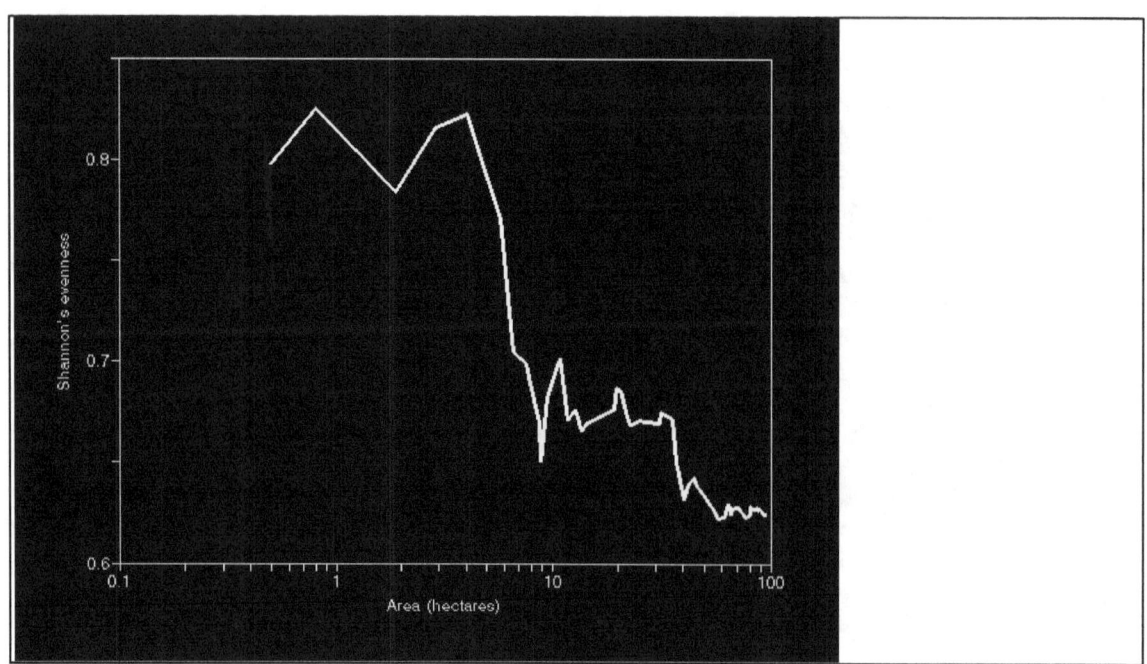

9.4 Discussion

The generally low abundances of most of the species of starfish at Heron Reef precluded the use of quadrats in general sampling. Because the traverse method will miss many cryptic individuals and provide only an approximate area measurement, the species diversity and species accumulation figures are only approximate. It would appear from Table 4.1 that most species occurred at a density that was less than one individual per hectare, with many species being far less abundant. It should be noted that traverse sampling will underestimate the density of all cryptic species, and will also fail to detect species that are both rare and cryptic.

McGuiness (1984) suggested that the use of species : (log) area or (log) species : (log) area for the display of the species : area relationship should be based on the underlying relative abundances of the species. The slope of the species : (log) area relationship, the slope of the (log) relative abundance : rank abundance relationship and Shannon's Evenness index are all indices of diversity. These allow a direct comparison to be made between different assemblages. Not only do these indices consider the number of species, they also express the inherent range of abundance between most common and least common within the assemblage (Connor and McCoy, 1979; Connor and Simberloff, 1979; Connor *et al.*, 1983).

Figures 9.1a,b show that the four most common species account for 70% of the total number of individuals in this assemblage. However, even *Echinaster luzonicus*, the most abundant species, had an average density of only 16 specimens per hectare. Of the 24 species of asteroid that occurred in the traverse samples, five species occurred only once. Presumably the species which were not found during this study, but which are known from the locality, occur with even less frequency than these five. Less than ten specimens of each of another six species were located on the intertidal traverses. Hence, 11 of the 24 species are regarded as very rare. Less than 25 specimens of another three species were found and these are regarded as rare. Thus a majority of the asteroid species found at Heron Reef are rare or very rare.

The slope of the regression line in Figure 9.1b is a measure of the diversity of this asteroid assemblage. The steeper the line the greater the range of relative abundance within a certain group of species. The less equal the relative abundances, the lower the diversity as measured by most diversity indices. Community studies often show a log-normal relationship in relative abundance, in which most species occur with close to the average abundance (Pielou, 1981). This assemblage of coral-reef asteroids does not clearly demonstrate this relationship, but this result may be attributable to an inadequate number of both species and individuals in the present study. The order of the species in Figures 9.1a,b is that of numerical abundance. If biomass or some other parameter was chosen as a measure of abundance, then the order of the species may change but the slope of the regression line might not alter greatly.

Figures 9.2a,b illustrate the species : area curve for the Heron Reef asteroid assemblage. The slope of the (log) area regression line is independent of the units used to measure area. Whether they be square metres or hectares, providing the habitat continues, the species will accumulate at a rate determined only by the relative abundances of the species in the assemblage. If there is some finite species pool which obviously cannot be exceeded, then the curve will become asymptotic.

The pronounced dips in Figures 9.3a,b are a result of small scale patchiness in the distribution of *Echinaster luzonicus* and *Disasterina abnormalis*. After continued sampling, the effect of this high localised abundance was rendered insignificant in the total diversity.

Figures 9.1a to 9.3b all relate to the one ecological parameter, namely the relative abundances of the species within this assemblage. This will determine the rate at which the species accumulate in a species : area curve, as well as the diversity as measured by most diversity indices.

The richness of the coral-reef asteroid assemblage at Heron Reef is unable to be compared directly with that of other coral-reef asteroid assemblages either on the Great Barrier Reef or elsewhere. This is because the extent of sampling has not been quantified in the majority of biogeographical studies. Because the area sampled determines the number of species in a sample of any assemblage (Fisher, Corbet and Williams (1943), the large number of species found at Heron Reef may be a result of the intensive sampling. Even so, it would appear from the linearity of Figures 9.2b that additional species of starfish occur intertidally at Heron Reef, but these species are either extremely rare or cryptic.

It is apparent that Heron Reef carries a rich and diverse asteroid fauna, 24 species belonging to six families having been found intertidally in 120 hectares during this study. The linearity of the species : (log) area relationship for the intertidal asteroid assemblage at Heron Reef indicates that additional species are still to be found. Indeed, *Mithrodia clavigera* was located subsequent to the traverses and Endean (1956) found three species (*Acanthaster planci*, *Ophidiaster watsoni* and *Anseropoda rosacea*) in the area of the traverses that were not found during the current study.

CHAPTER 10

GENERAL DISCUSSION

Population density, size-frequency and reproductive data on an assemblage of shallow water, coral-reef starfish (Asteroidea) were gathered over five years at Heron Reef. Heron Reef, which is located near the southern end of the Great Barrier Reef, has not been known to carry an outbreak of the crown-of-thorns starfish (*Acanthaster planci*) and its coral cover is well developed. While there has been detailed study of the starfish assemblages on some reefs that have recently undergone *Acanthaster planci* population outbreaks (Yamaguchi, 1975 b; 1977 a), the composition of these assemblages may well be different from pre-outbreak assemblages.

Abundance, size-frequency and reproductive data were collected by means of intertidal traverses which ran between the cay and the reef crest (0.5 to 2 kilometres apart) and also between two points both on the reef crest (0.5 to 6 kilometres apart). Most traverses included both reef flat and reef crest zones, and all exposed starfish within a 4 meter width were collected. A selection of large and small, dead coral slabs occurring on these traverses were overturned and cryptic specimens located beneath these slabs were collected also. In total, 72 intertidal traverses were conducted covering an area of approximately 120 hectares (1.2 square kilometres). Cryptic species were also sampled using metre square quadrats in particular areas where previous traverse sampling had shown that starfish abundance was relatively high. Subtidal specimens of starfish were collected on the reef slope and off-reef floor by the use of SCUBA.

Of the 25 starfish species found on Heron Reef, *Asteropsis carinifera*, *Dactylosaster cylindricus*, *Fromia milleporella*, *Linckia laevigata*, *Nardoa novaecaledoniae*, *N. pauciforis*, *Ophidiaster confertus*, *O. granifer*, *O. lioderma*, *O.*

robillardi, *Asterina anomala*, *A. burtoni*, *Disasterina abnormalis*, *D. leptalacantha*, *Tegulaster emburyi*, *Mithrodia clavigera* and *Coscinasterias calamaria* were located only in intertidal regions. *Linckia guildingii*, *L. multifora* and *Echinaster luzonicus* were found predominantly in intertidal regions but some specimens were located subtidally. *Culcita novaeguineae*, *Acanthaster planci*, *Fromia elegans*, *Gomophia egyptiaca* and *Neoferdina cumingi* were located predominantly in subtidal habitats, but are known to occur intertidally. *Culcita novaeguineae* seemed to mainly inhabit the deeper coral pools adjacent to the lagoon. The low occurrence of *Culcita novaeguineae* on the intertidal traverses is because the traverses avoided this slightly deeper-water habitat. While *Culcita novaeguineae*, *Fromia elegans*, *Gomophia egyptiaca*, *Linckia multifora* and *Echinaster luzonicus* were sometimes found at the base of the reef slope, they were never observed on the sea floor away from the reef. There are no published records of these species from the off-reef floor zone (see Clark and Rowe, 1971).

"Reef" echinoderm species were separated from "mainland" species on the basis of their habitat requirements by Endean (1956) who discussed the biogeographical relationships of Great Barrier Reef species. With the exception of *Ophidiaster confertus* and *Coscinasterias calamaria*, which are essentially temperate species, 23 asteroid species found at Heron Reef can be regarded as coral-reef species and their distribution differs from species such as *Astropecten polyacanthus*, *Iconaster longimanus*, *Pentaceraster regulus*, *Leiaster leachi*, *Nardoa rosea*, *Ophidiaster armatus*, *Tamaria megaloplax* and *Echinaster stereosomus*. These latter species appear to be predominantly off-reef, sea-floor species that are widely distributed throughout the shallow waters of tropical and sub-tropical Queensland. The predominantly reefal distribution of the long-spined, corallivorous species, *Acanthaster planci*, contrasts with that of its generally deeper water, short-spined, molluscivorous relative, *A. brevispinus*. Only small fissiparous specimens of *Coscinasterias calamaria* were located on Heron Reef. Large adults of this and other forcipulatid species are predators in temperate communities. Both *Ophidiaster confertus* and *Coscinasterias calamaria* appear to be predominantly temperate species that occur in Australian mainland waters but which have extended their ranges to reefs at the southern end of the Great Barrier Reef.

The finding of *Iconaster longimanus*, *Asteropsis carinifera*, *Dactylosaster cylindricus*, *Fromia elegans*, *Linckia multifora*, *Ophidiaster armatus*, *Ophidiaster lioderma*, *Ophidiaster robillardi*, *Tamaria megaloplax*, *Asterina anomala*, *Disasterina abnormalis*, *Tegulaster emburyi*, *Mithrodia clavigera*, *Echinaster stereosomus* and *Coscinasterias calamaria* represent new records for Heron Reef. In some cases these represent new records for the Great Barrier Reef, and in other cases known ranges on the Great Barrier Reef have been considerably extended. This study has also provided the first record of the predominantly temperate species, *Coscinasterias calamaria* on the Great Barrier Reef.

The distinguishing characteristic of coral-reef species of starfish is their possession of a spatial distribution that never extends into the deeper parts of the off-reef floor zone. Such a spatial distribution would preclude between-reef migration by post-larval stages of these species. It is not known why some species of starfish are essentially restricted to coral reefs, but it is likely that such species would differ in their physiological and / or ecological requirements from species that occur elsewhere. While the intertidal region of a coral reef undergoes both temperature and salinity fluctuations (Maxwell, 1968), a substrate of coral sand and rubble (aragonite not calcite) would ensure complete carbonate saturation of the waters and hence the waters would be well buffered against pH changes. Some species of starfish that occur exclusively in association with coral reefs may have narrow pH tolerances. Other species may have evolved interdependencies that involve settlement or survival conditions that are only present within the coral reef ecosystem. Likewise, with respect to the coral reef ecosystem itself, it might be expected that species that occur predominantly in one of the major zones of a coral reef (e.g. the reef flat) would differ in their physiological and / or ecological requirements from species that occur in several of these zones. For example, they might differ in their degree of tolerance to sub-aerial exposure at low tide or in their biotic associations.

Patches of localised high density were observed within the populations of some of the smaller-bodied species of coral-reef starfish that were studied. However, each of these patches appeared to be restricted to a very small area. For example, the small-bodied starfish *Disasterina abnormalis* occurred at an average density of over eight individuals per square metre at one location on the northern reef crest but 100 metres away (still on the reef crest) its density was less than one individual per square metre. This region of high density of *Disasterina abnormalis* appeared to be confined to a narrow strip behind a rubble bank and this species was not found on 25 of the 72 traverses that were made. In this region, *Disasterina abnormalis* was highly clumped (at the metre square scale) in one sampling period and randomly distributed in another sampling period.

Echinaster luzonicus was the most abundant starfish found on the intertidal traverses and *Linckia multifora* was the next in order of decreasing abundance. Both of these small-bodied species were found in relatively high numbers in some regions of the reef crest. The large-bodied starfish *Linckia laevigata* was third in order of decreasing abundance on the traverses but its maximum density did not approach that of either of the preceding species anywhere at Heron Reef. The density of *Linckia laevigata* at Heron Reef appeared to be low compared with its density on reefs that are known to have carried an outbreak of *Acanthaster planci* (Laxton, 1974; Yamaguchi, 1977 a; Thompson and Thompson, 1982; Klumpp and Pulfrich, 1989). Laxton (1974) suggested that *Linckia laevigata* may either increase its numbers or extend its range following outbreaks of *Acanthaster planci*. *Disasterina abnormalis* was fourth in order of decreasing abundance and occurred at the highest local density of any species of starfish during this study.

The intertidal traverses made during this study covered an area of 125 hectares. Over 1400 individuals of *Echinaster luzonicus* were located and over 100 individuals of each of another 8 species were located. However, fewer than 25 individuals of each of the remaining 15 species were located. The low starfish density found at most locations on Heron Reef contrasts markedly with the high densities recorded for asteroids of temperate communities (Loosanoff, 1961; 1964; Mauzey *et al*, 1968; Menge, 1975; Dayton *et al*, 1977; Birkeland *et al*, 1982; Stevenson, 1992).

Traverse sampling resulted in the location of a total of 24 species of intertidal starfish. For 10 of these species, a sufficient number of individuals was obtained for reproductive analysis and for 7 of these species size-frequency variation was examined over different sampling periods. Traverse sampling enabled data to be gathered on a large spatial scale (125 hectares) which facilitated both the collection of sufficient specimens for reproductive and size-frequency analysis as well as the determination of large scale non-randomness in the spatial distribution of these species.

While the intertidal traverse data did not allow small-scale analysis of either spatial or temporal abundance variation, the starfish assemblage at Heron Reef clearly embraces a highly diverse and spatially heterogeneous group of species. Individuals of each species were extremely non-random (clumped) in their spatial distribution. Only *Echinaster luzonicus* was sufficiently abundant and widespread to be found on all but three of the traverses. *Linckia laevigata* and *Nardoa novaecaledoniae* were not located on 10, *Nardoa pauciforis* was not located on 19, *Linckia multifora* was not located on 22, *Disasterina abnormalis* was not located on 25, *Asterina burtoni* was not located on 26 and *Linckia guildingii* was not located on 34 of the 72 traverses made. Representatives of the remaining species were not found on the majority of these intertidal traverses.

With the exception of *Echinaster luzonicus*, the abundance distributions of all of the species had a modal traverse density of zero individuals per hectare. This indicated that, with the exception of *Echinaster luzonicus*, each coral-reef starfish species was not represented on a large number of the traverses. The more common of these species possessed a bimodal abundance distribution which indicated that they were non-random (patchy) in their spatial distribution. For these species, there were many traverses where both zero and a relatively large number of individuals per hectare were recorded and very few traverses where intermediate (mean) densities occurred.

Table 4.1 lists the mean density per hectare and the variation that occurred in the mean density of each species between traverses. In all species the standard deviation was greater than the mean density. These data together with the bimodal population distribution data (Figures 4.2 to 4.12) indicate that large scale aggregation occurs in all of the species with the possible exception of *Echinaster luzonicus*. A stratified-random sampling procedure, using multiple belt transects would have allowed a detailed comparison of starfish abundances between different habitats. However, when used on a reef that has low general starfish abundance, such a sampling method would not have located a sufficient number of individuals in the limited time available for field studies at Heron Reef to permit a statistically valid size-frequency and reproductive analysis.

A mode in the abundance distribution was recorded at between three and 10 individuals / hectare in six species (rank 1 - 6) and at between one and three individuals / hectare in another six species (rank 7 - 12). The remaining twelve species (rank 13 - 20) were encountered so infrequently that the only mode in the abundance distribution of each species was at zero individuals per hectare. Five species were sufficiently uncommon (rank 20) to be encountered on only one intertidal traverse during the entire study.

Culcita novaeguineae, *Fromia elegans*, *Gomophia egyptiaca* and *Nardoa rosea* were encountered much more frequently in sub-tidal traverses than they were on intertidal traverses. *Disasterina leptalacantha* was recorded more frequently at Heron Island by Endean (1957) than it was in this study, but there may have been confusion between the two similar congeneric species in the earlier study. Similarly the ecological distinction between *Asterina anomala* and *Asterina burtoni* is unclear. The observed variation in the abundance of *Asterina burtoni* at Heron Reef is consistent with the results of Achituv and Sher (1991), but the mode of reproduction appears to be different.

The very small and highly cryptic species *Disasterina abnormalis* occurred periodically with high abundance at one location on the inner reef crest. It was possible to sample this species in this localised habitat by means of metre square quadrat sampling (Table 4.2). The data obtained do not represent the abundance of this species generally, but serve to illustrate clearly the enormous spatial and temporal variation that occurs in the population distributions of this opportunistic species.

Although the diets of the coral-reef starfish species encountered were not studied in detail, many of them appeared to feed on epibenthic felt. In every coral reef zone, some species sought no refuge and occurred in exposed situations. Clear examples of niche (dietary or microhabitat) specialisation are known only for *Culcita novaeguineae* and the predominantly subtidal species *Acanthaster planci* both of which feed primarily on corals. Competitive interactions were not studied, but many species occurred at a sufficiently low density that they may not be resource limited.

Because of the patchy nature of the spatial distributions of all of the coral-reef asteroid species, size-frequency analysis over multiple sampling periods (Tables 5.1 to 5.12 and Figures 5.1a to 5.10d) was considered the most appropriate means of establishing the existence of population stability. Obvious changes in abundance due to either sexual or asexual recruitment, and significant changes in mean individual size were observed in the populations of *Linckia multifora*, *Disasterina abnormalis* and *Echinaster luzonicus* (Table 8.1 and Figures 8.1a to 8.3c). While some recruitment and some change in abundance was noticed in both *Ophidiaster granifer* (parthenogenetic) and *Asterina burtoni* (hermaphroditic), no significant change occurred in the mean individual size of either species. *Linckia guildingii*, *Linckia laevigata*, *Nardoa novaecaledoniae* and *Nardoa pauciforis* exhibited only small changes in mean individual size and these species did not fluctuate greatly in abundance during the period of study. Also, the population structure of these species appeared to be adult dominated and juveniles were encountered only rarely.

The remaining species were not found in sufficient numbers for meaningful statistical analysis of size-frequency data. Their populations were sparse and juveniles were not encountered except for one specimen each of *Culcita novaeguineae*, *Fromia elegans* and *Gomophia egyptiaca*. Their populations appeared to be adult dominated. Juveniles of *Culcita novaeguineae* and *Fromia elegans* were not encountered subtidally despite the existence of a subtidal population of adults. One juvenile of *Acanthaster planci* was located at the base of the reef slope.

Culcita novaeguineae and many other coral-reef starfish species were not encountered in sufficient numbers to warrant an examination of their population stability. The study of Laxton (1974) appeared to show a greater abundance of *Linckia laevigata* on the reef flat at Heron Reef than was observed in this study. Laxton suggested that this species may vary its distribution range following outbreaks of *Acanthaster planci*. It is possible that large-bodied species of starfish, such as *Linckia laevigata*, undergo large scale aggregation behaviour but the limited duration of this study precluded examination of such long period fluctuations.

Grassle (1973), Sale and Dybdahl (1975), Talbot *et al.* (1978) and Hutchings (1981) all found that most coral-reef species are rare. Endean and Cameron (1990 a) mention that the high incidence of rare species in the coral-reef community contributes markedly to species diversity. Some of the rarer species of coral-reef starfish are known from only a few specimens and their low-density populations defy our normal understanding of population dynamics and reproductive strategies. It is not clear how these species survive or whether their populations are predator, resource or recruitment limited. Species such as *Tosia queenslandensis*, *Ophidiaster lioderma* and *Tegulaster emburyi* have always been considered rare throughout their geographical range. Although nothing is known of their reproductive cycles, if they are truly rare and valid "biological" species, then they might be expected to exhibit mechanisms such as population aggregation, asexual reproduction, parthenogenesis or hermaphroditism that would facilitate their persistence at low population densities.

Inter-coelomic injection with the hormone 1-methyl adenine was used to determine the sex ratio, reproductive maturity and type of larval development of several of the species. It can be seen from Tables 6.1 to 6.8 and Figures 6.1 to 6.8 that eight of the more common species appeared to demonstrate an annual sexual reproductive cycle. *Disasterina abnormalis* possessed small (non-yolky) sticky eggs that adhered to the substrate immediately following their release from the gonopores. Small juveniles of *Disasterina abnormalis* were relatively common in one highly localised area at Heron Reef, but high settlement was not observed in any of the other species. The remaining seven species possessed eggs that dispersed and underwent either planktotrophic or lecithotrophic larval development. No species were observed to brood larvae.

Culcita novaeguineae, *Acanthaster planci*, *Linckia guildingii* and *Linckia laevigata* were observed releasing eggs that contained little yolk and underwent planktotrophic development. *Fromia elegans*, *Gomophia egyptiaca*, *Nardoa novaecaledoniae*, *Nardoa pauciforis*, *Ophidiaster granifer* and *Echinaster luzonicus* were observed releasing eggs that contained large amounts of yolk and underwent lecithotrophic development. Specimens of both *Linckia multifora* and *Asterina burtoni* were injected regularly, but did not release gametes during the entire study.

Vance (1973) and Yamaguchi (1973 a, 1973 b, 1977 b) suggested that lecithotrophic development is an adaptation to high predation or starvation of larva because with this development the length of larval life can be shorter than with planktotrophic development. On Heron Reef, and possibly the Great Barrier Reef in general, where many reefs exist in relatively close proximity, lecithotrophic genera such as *Nardoa*, *Fromia* and *Echinaster* might be expected to be better represented than they are on widely scattered atolls. At Heron Reef, the larger-bodied species namely, *Culcita novaeguineae*, *Acanthaster planci*, *Linckia guildingii* and *Linckia laevigata* all liberated dispersing, small eggs that underwent planktotrophic development while the smaller-bodied species, together with *Nardoa novaecaledoniae* and *Nardoa pauciforis* (both intermediate in body size), all liberated larger eggs that underwent lecithotrophic development. The small, sticky eggs of *Disasterina abnormalis* resulted in high localised settlement and this strategy appeared to be unique amongst the starfish species that were studied at Heron Reef.

In addition to the species that demonstrated a sexual reproductive cycle, *Linckia guildingii*, *Linckia multifora*, *Ophidiaster robillardi* and *Echinaster luzonicus* reproduced asexually and exhibited comet stages while *Asterina anomala* and *Coscinasterias calamaria* reproduced asexually by binary fission. All small specimens of these species exhibited the characteristics of either autotomous propagation (see Rideout, 1978) or binary fission. While all of the arms might look quite similar in some small individuals of autotomous species, the original arm from which the others regenerated was always apparent following closer examination. All specimens of fissiparous species showed signs of recent binary fission.

While specimens of both *Linckia guildingii* and *Echinaster luzonicus* were observed releasing gametes in response to injection with 1-methyl adenine, no sexually-propagated juveniles were observed in the populations of any species that reproduced asexually. With the exception of *Linckia guildingii*, large bodied species of coral-reef starfish do not appear to have a small scale (low dispersion) reproductive strategy. This could indicate that survival of offspring is more likely away from adult populations. The advantages of a high dispersion reproductive strategy must be balanced against the high dispersive loss resulting from the relative isolation of reefs of the Great Barrier Reef and elsewhere.

Linckia multifora and *Echinaster luzonicus* were the only asexually reproducing species in which high rates of autotomy were observed and the location of comet stages and adults in various stages of regeneration is evidence of relatively high asexual recruitment. These three species had the highest localised abundances of any of the coral-reef starfish but also had highly patchy spatial distributions. The remaining species never occurred at densities comparable with these species even though the average density of *Linckia laevigata* was higher than the average density of *Disasterina abnormalis*. While comet stages and adults in various stages of regeneration were observed in *Linckia guildingii*, this species did not show evidence of high asexual recruitment.

With the exception of *Disasterina abnormalis* (see Chapter 6), all the species of starfish at Heron Reef either possessed a planktonic dispersive larval phase or were not observed to reproduce sexually. The largest-bodied persistent species released planktotrophic eggs while the opportunist species were either lecithotrophic, hermaphroditic (*Asterina burtoni*), parthenogenetic (*Ophidiaster granifer*) or solely asexually reproducing (*Linckia multifora*). *Nardoa novaecaledoniae*, *Nardoa pauciforis* and *Gomophia egyptiaca* would appear to be of intermediate position and the taxonomic position of *Asterina anomala* is unclear.

All of the large-bodied species studied liberated either eggs or sperm directly into the water column and fertilisation was external. While possible pairing was observed in crowded aquaria (following injection with 1-methyl adenine), no species were observed mating in the field as has been recorded by Run, Chen, Chang and Chia (1988) for the tropical species *Archaster typicus*. Slattery and Bosch (1993) also recorded mating behaviour in an Antarctic species of starfish.

Ormond *et al.* (1973) discussed the consequences of spawning aggregations of *Acanthaster* and suggested that the increased proximity of adult starfish may enhance the chances of fertilisation, especially if synchronous spawning takes place. It was suggested by Lucas (1984) that a conspecific stimulus would induce synchronous spawning in *Acanthaster planci* and a delayed spawning activity in dispersed individuals of

Acanthaster planci was observed by Okaji (1991). It was suggested that this delay reflected less frequent stimulus from conspecifics in dispersed populations compared with aggregated populations and that synchronous spawning induced by such stimulus would lead to higher rates of fertilisation when the animals formed an aggregation. Evidence of the existence of sexual pheromones in starfish was presented by Miller (1989).

The effect of sperm dilution, adult aggregation and synchronous spawning upon the fertilisation of sea-urchin eggs was reported by Pennington (1985). Pennington concluded that significant fertilisation occurred only when spawning individuals are closer than a few metres. The consequences of water mixing and sperm dilution for species that undergo external fertilisation were discussed by Denny and Shibata (1989) who found that only a small fraction of ova were fertilised other than in densely packed arrays. They commented that the low effectiveness of external fertilisation may change the way one views the planktonic portion of such life cycles and suggested that this could serve as a potent selective factor. For the rarer sexually reproducing species, it is apparent that aggregation resulting in the occurrence of an opposite sexed conspecific within the effective fertilisation distance is a condition precedent to successful reproduction. The degree of reproductive success may be strongly dependent on just how close the rare spawning individuals are to each other. While the results of Babcock and Mundy (1992) appear inconsistent with these previous studies, the population density and degree of adult aggregation would be highly relevant factors for both the synchrony of spawning and the level of egg fertilisation in externally fertilising dioecious species. If a low density starfish population is highly dispersed then the degree of egg fertilisation would be much lower than if aggregation occurred.

The above factors influence recruitment as do many other factors such as dispersion loss (Atkinson *et al*, 1982; Dight *et al.*, 1990 a, b; Black and Moran, 1991; Wolanski, 1993) and starvation of larvae (Birkeland, 1982; Olsen, 1987). These factors, together with the mortality of juveniles prior to first reproduction (Endean, 1977; 1982; McCallum et al, 1989), might result in this assemblage being recruitment limited as suggested for certain species of coral-reef fish by Doherty (1982). If the process of recruitment is completed when an organism enters the breeding population, then a species could be regarded as recruitment limited if mortality of its larvae or juveniles was sufficiently great to maintain adult populations at a low density. This may occur as a result of either low egg fertilisation or high mortality of larvae or juveniles.

On reefs such as Heron Reef that have low adult starfish abundance, predation of adult starfish appears to be a rare event and was not studied because of logistic constraints. While the giant triton (*Charonia tritonis*) is a voracious predator of large juvenile and adult starfish (Endean, 1969; Pearson and Endean, 1969), no specimens of this species were observed at Heron Reef either subtidally or on intertidal traverses during the entire study. The giant triton is cryptic and it is extremely difficult to survey the population density of this predator. It is likely that there are other predators of coral-reef starfish, particularly fishes. Other predators (see Endean and Cameron, 1990 b) have been found for *Acanthaster planci*. If starfish populations are stable then mortality (including lethal predation) will match recruitment which appeared to be extremely low in the populations of large bodied coral-reef starfish. If starfish populations are maintained at a low adult density, then predation on pre-adults could be a major factor in controlling the assemblage.

An increase in anti-predatory structures with decreasing latitude was found by Vermeij (1978) and Blake (1983) suggested the existence of a similar pattern in sea stars. Pearson and Endean (1969) and McCallum *et al.* (1989) reported a high incidence of sub-lethal predation in populations of *Acanthaster planci*. Blake (1983) commented that the asteroid fauna of the Indo-West Pacific are dominated by the order Valvatida and members of this order have the best developed anti-predatory devices. Yamaguchi (1975 b) commented on the difference between adult and juvenile asteroid habits and suggested that the heavy armour of exposed adult asteroids might reflect heavy predation pressure.

In addition to the protection afforded by structural features, many species of starfish are protected from generalist predation by the possession of skin toxins (Riccio *et al.*, 1982, 1985; Gorshkov *et al.*, 1982; Minale *et al.*, 1984; Narita *et al.*, 1984; Noguchi *et al.*, 1985 a,b; Miyazawa *et al.*, 1985; 1987; Kicha *et al.*, 1985; Shiomi *et al.*, 1988; Shiomi *et al.*, 1990; Zagalsky *et al.*, 1989; Iorizzi *et al.*, 1991; Bruno *et al.*, 1993; Casapullo *et al.*, 1993). These skin toxins have been shown to be toxic to some fish species (Rideout, 1975). The role of echinoderm toxins as a defence against predation has been discussed extensively (Bakus, 1974; Green, 1977). Cameron and Endean (1982) discussed the role of venomous devices and toxins as defences against predation and Endean and Cameron (1990 a) have noted that persisters are often toxic. There is little information available on the toxicity of juvenile starfish to potential predators. Eggs and juveniles of *Acanthaster planci* are known to carry toxins. It has been proposed that the production of toxins for defence incurs an energy cost which is balanced against the probability of mortality (Eckardt, 1974) but in some species, toxins might be metabolic by-products that incur no energy cost in their synthesis.

In some groups of starfish behavioural mechanisms are used as defences against predation and Blake (1983) suggested that both *Luidia* and *Astropecten* have broad open ambulacral furrows because they were predators on active solitary forms where increased skeletal mobility was essential. Because both these active, hunting genera live on and within unconsolidated sediment they avoid predation by burrowing which is facilitated by the paxillose nature of their aboral surface.

Another behavioural defence possessed by asteroids is the autotomy of arms. Of the coral-reef starfish studied, *Linckia guildingii*, *Linckia multifora*, *Ophidiaster robillardi* and *Echinaster luzonicus* are capable of regenerating a complete individual from the distal section of one arm. These autotomous species were extremely aggregated in their spatial distribution, suggesting that population growth occurs with little dispersal of individuals.

In species of starfish that do not reproduce by autotomy, specimens are often observed in various stages of regeneration following loss of one or more arms. McCallum *et al* (1989) reported that 40% of the adult individuals in a population of *Acanthaster planci* showed signs of arm regeneration. Cameron and Endean (1982) suggested that autotomy is an adaptation to predation and Birkeland *et al* (1982) observed autotomy in their study of asteroid predatory interactions. At Heron Reef, many individuals were observed in various stages of regeneration following autotomy of one or more limbs. A number of tropical asteroids are known to undergo regular autotomy (Rideout, 1978; Yamaguchi, 1975 b) and Blake (1983) commented that interpretation of the skeleton can be difficult as it has more than one function and protection against predation can be accomplished by many mechanisms (e.g. Bullock, 1953; Feder, 1963; Mauzey *et al.*, 1968; Ansell, 1969; Birkeland, 1974; Phillips, 1976; Dayton *et al.*, 1977; Jost, 1979; Schmitt, 1982; Stevenson, 1992; Iwasaki, 1993).

In the species that reproduce by autotomy, it is not known to what extent the autotomisation of a limb is caused by physical disturbance such as predation. While direct predation was not observed, large individuals of the large-bodied species of starfish often had their arms intertwined with the substrate such that they were difficult to dislodge. In the large-bodied species that only reproduce sexually, parts of a limb and even one or two whole limbs were observed to be missing from some individuals. The existence of such behaviour together with the observations of missing arms in species that do not reproduce asexually, indicates that sub-lethal predation does occur. Whether it is significant in the regulation of the Heron Reef asteroid assemblage will depend on the age structures of the populations. Sub-lethal predation of adults will be especially important if a species is long lived.

This study has examined the population dynamics of both relatively common and relatively rare species of coral-reef starfish. Although some species were not sufficiently numerous to provide statistically satisfactory numbers of records, data were gathered on their habitat, size, spatial pattern and relative abundance. It is clear that the majority of species of intertidal starfish at Heron Reef were sufficiently uncommon to preclude small scale methods of population examination. There is considerable disagreement over the accuracy of large scale methods (manta tow) to examine subtidal populations of starfish (Fernandes, 1990; Fernandes *et al.*, 1990; Moran and De'ath, 1992 a,b). However, the determination of large scale, non-random variation in the distribution of any species is a condition precedent to the determination of its overall abundance. In the estimation of average density, methods of both sampling and analysis must adequately consider the high standard error of the mean. All conclusions must have due regard to the bimodality and skewness of the abundance distributions of starfish.

Some species, namely *Disasterina abnormalis*, *Asterina burtoni*, *Ophidiaster granifer*, *Linckia multifora* and *Echinaster luzonicus*, could be regarded as opportunist species as they were characterised by possessing relatively abundant populations with relatively large fluctuations in mean individual size. These invariably small-bodied species demonstrated all of the typical opportunist characteristics which are short life, high recruitment and high mortality (see Endean and Cameron, 1990 a).

Other species, namely *Culcita novaeguineae*, *Linckia laevigata*, *Linckia guildingii*, *Nardoa novaecaledoniae* and *Nardoa pauciforis* could be regarded as persistent species and were characterised by less abundant populations with relatively smaller fluctuations in mean individual size. These invariably medium to large bodied species demonstrated all of the typical persister characteristics which are long life, low recruitment and low mortality. A large proportion of coral-reef starfish were sufficiently uncommon to preclude any analysis of either their abundance or size distributions. Apart from the knowledge that they remained rare through the study period of 5 years, little is known of their natural history. Because of their extreme rarity, which is a characteristic of persisters, they might be placed in this category pending further investigation. Of the 25 intertidal species of starfish, five species (20 percent) were characteristic opportunist coral-reef species and 18 species (72 percent) were characteristic persister coral-reef species (stable abundance and size distribution or remained uncommon throughout study). Only two species (8 percent), namely *Ophidiaster confertus* and *Coscinasterias calamaria* were sub-tropical, rocky-reef (mainland) species that had extended their ranges to embrace the southernmost reefs of the Great Barrier Reef.

The longevity of a species is determined by the relative probability of juvenile and adult survivorship. In the simplest case, if the probability of a sexually mature organism's survival from one reproductive season to the next is greater than the probability of one of the offspring reaching sexual maturity, then the species will exhibit iteroparity (see Cole, 1954; Murphy, 1968; Goodman, 1974;

Stearns, 1977; Roff, 1981; Ebert, 1982). Although neither predation nor mortality was observed during this study, both low adult mortality and relative longevity can be inferred from the stability of the size-frequency distributions of the persistent species studied. This contrasts with the large population fluctuations and instability of the population structure of the opportunist species studied.

Most marine benthic invertebrates have a high energy cost associated with reproduction (Mileikovsky, 1971).
Under differing selection pressures, it has been suggested that long life can be associated with either variable recruitment (Sterns, 1977) or fixed low recruitment (Charnov and Schaffer, 1973; Schaffer, 1974; Ebert, 1982). McCallum (1987) and McCallum *et al*. (1989) have suggested that *Acanthaster planci* is recruitment limited by juvenile and sub-adult predation.

A model relating to our perception of the life history of all organisms, referred to as r- versus K- strategy, was reviewed by Stearns (1977). The different survival characteristics in the model were thought to have evolved in response to specific types of environments (Murphy, 1968; Hairston, Tinkle and Wilbur, 1970). The spectrum of existing life history attributes, apparent in any community study (see e.g. Menge, 1975; Vance, 1973), was considered to represent many points on a continuum between the conceptually ideal r- strategists and K- strategists.

It has been suggested that the dispersal stage of a population spreads the risk of local extinction in space and time (Den Boer, 1971; Scheltema, 1971; Strathmann, 1974). Opportunists survive by being able to colonise regions quickly following disturbance. In this regard, an important distinction must be made between equilibrium and non-equilibrium populations in terms of adaptive characteristics (Caswell, 1982; Ebert, 1985). High spatial and temporal variation in population size seems to characterise the typical opportunists.

The degree of spatial and temporal stability in the population of a species determines its position on a theoretical opportunist - persister continuum. Each species was viewed in this context and a basic dichotomy was observed. Because it does not require presumptions of carrying capacity, and inferences about competition, the opportunist / persister model of Endean and Cameron (1990 a) seems to best describe this low density assemblage of coral-reef starfish. Stable ecosystems should be characterised by small fluctuations of their component species. However it is clear that the apparent stability or instability of any biological system is dependent not only on the spatial and temporal scales of observation (Bradbury and Reichelt, 1982; Sale, 1984; Weiss, 1969) but also on the particular subset of species that is examined.

The observed level of numerical and size-frequency stability in the persistent coral-reef asteroid species is consistent with a model of community equilibrium. It is clear that mortality, dispersion, larval survival and settlement phenomena did not result in widely varying size structures or greatly differing adult numbers from one year to the next over a period of 5 years. The vast majority of species of coral-reef starfish in the assemblage studied were characterised by continuing low abundance. It would appear that when a rare, large-bodied starfish is established in its adult population, it is likely to be long lived. *Acanthaster planci* is a member of this coral-reef starfish assemblage and Cameron (1977) has suggested that only when the coral reef ecosystem is drastically altered can such a rare and long-lived carnivore undergo population outbreaks. This restriction may also apply to other persistent species in the coral-reef starfish assemblage.

Factors such as high gamete dilution (Rothschild and Swann, 1951; Pennington, 1985; Denny and Shibata, 1989; Epel, 1991), as well as basically unpredictable environmental factors such as larval mortality and enormous potential larval dispersion can affect the number of larvae reaching a reef. Because the area of coral reef in the Great Barrier Reef region is relatively small compared with the area of sea surface in the region, the probability of a planktonic starfish larva reaching a coral reef is quite low. Also, if predation on post-settlement juveniles is intense then recruitment will be low. In low density starfish populations, the aggregation of adults prior to spawning may be essential to the reproductive success of a rare species. Because successful recruitment implies that post-settlement juveniles must survive to enter the breeding population, predation on juveniles as well as sub-lethal predation of adults (when loss of gonad affects fecundity) are both forms of recruitment limitation.

The results presented in this study are in accord with the hypothesis of Endean and Cameron (1990 a) that complex, high diversity assemblages of coral-reef animals are characterised by a predominance of rare, long-lived species with relatively constant population sizes and size structures and a minority of relatively common, short-lived opportunistic species characterised by fluctuating population sizes and size structures.

REFERENCES CITED

Abbott,I. 1983.
The meaning of z in species / area regressions and the study of species turnover in island biogeography.
Oikos 41: 385-390.

Achituv,Y. 1972.
The genital cycle of *Asterina burtoni* Gray [Asteroidea] from the Gulf of Elat, Red Sea.
Cah. Biol. Mar. 14(4): 547-553.

Achituv,Y. and Z.Malik. 1985.
The spermatozoa of the fissiparous starfish, *Asterina burtoni*.
Int. J. Invert. Repro. Devel. 8: 67-72.

Achituv,Y. and E.Sher. 1991.
Sexual reproduction and fission in the sea star *Asterina burtoni* from the Mediterranean coast of Israel.
Bull. mar. sci. 48: 670-678.

Ansell,A.D. 1969.
Defensive adaptations to predation in the mollusca.
Mar. Biol. Assoc. India 3: 487-512.

Antonelli,P.L. and N.D.Kazarinoff. 1988.
Modelling density-dependent aggregation and reproduction in certain terrestrial and marine ecosystems: A comparative study. Ecol. modelling 41: 219-228.

Atkinson,M.J., S.V.Smith and E.D.Stroup, 1982.
Circulation in Enewetak Atoll lagoon.
Proc. 4th Int. Coral Reef Symp. 1: 335-338.

Babcock,R.C. and C.N.Mundy. 1992.
Reproductive biology, spawning and field fertilisation rates of *Acanthaster planci*.
Aust. J. Mar. Freshwater Res. 43: 525-534.

Baker,A.N. and L.M.Marsh. 1974.
The rediscovery of *Halityle regularis* Fisher [Echinodermata, Asteroidea]. Rec. W.A. Mus. 4(2): 107-116.

Bakus,G.J. 1974.
Toxicity in holothurians: a geographical pattern.
Biotropica 6(4): 229-236.

Barker,M.F. 1977.
Observations on the settlement of the brachiolaria larvae of *Stichaster australis* (Verrill) and *Coscinasterias calamaria* (Gray) (Echinodermata: Asteroidea) in the laboratory and on the shore. J. exp. mar. Biol. Ecol. 30: 95-108.

Bennett,I. 1958.
Echinoderms from the Capricorn Group, Queensland, 23-27 S.
  Proc. Linn. Soc. N.S.W. 83: 375-376.

Benson,A.A., Patton,J.S. and C.E.Field. 1975.
Wax digestion in the Crown of Thorns starfish.
Comp. Biochem. Physiol. B. Comp. Biochem. 52(2): 339-340.

Benzie,J.A.H. and J.A.Stoddart. 1992.
Genetic structure of outbreaking and non-outbreaking crown-of-thorns starfish (*Acanthaster planci*) populations on the Great Barrier Reef. Mar. Biol. (Berlin) 112: 119-130.

Benzie,J.A.H. and J.A.Stoddart. 1992.
Genetic structure of crown-of-thorns starfish (*Acanthaster planci*) in Australia. Mar. Biol. (Berlin) 112: 631-639.

Birkeland,C. 1974.
Interactions between a seapen and seven of its predators.
Ecol. Monogr. 44: 211-232.

Birkeland,C. 1982.
Terrestrial runoff as a cause of outbreaks of *Acanthaster planci*. Mar. Biol. 69(2): 175-186.

Birkeland,C., Dayton, P.K. and N.A.Engstrom. 1982.
A stable system of predation on a holothurian by four asteroids and their top predator. Aust. Mus. Mem. 16: 175-189.

Black,K.P. 1993.
The relative importance of local retention and inter-reef dispersal of neutrally buoyant material on coral reefs. Coral reefs 12: 43-53.

Black,K.P. and P.J.Moran. 1991.
Influence of hydrodynamics on the passive dispersal and initial recruitment of larvae of *Acanthaster planci* (Echinodermata: Asteroidea) on the Great Barrier Reef. Mar. ecol. prog. ser. 69: 55-65.

Blake,D.B. 1979.
The affinities and origins of the crown-of-thorns sea star *Acanthaster* Gervais. J. Nat. Hist. 13: 303-314.

Blake,D.B. 1980.
Affinities of three small sea-star families.
J. Nat. Hist. 14: 163-182.

Blake,D.B. 1981.
A reassessment of the sea-star orders Valvatida and Spinulosida. J. Nat. Hist. 15: 375-394.

Blake,D.B. 1983.
Some biological controls on the distribution of shallow water sea stars (Asteroidea; Echinodermata).
Bull. Mar. Sci. 33: 703-712.

Blake,D.B. 1987.
A classification and phylogeny of post-Paleozoic sea stars (Asteroidea: Echinodermata).
J. nat. hist. 21: 481-528.

Blake,D.B. 1990.
Adaptive zones of the class Asteroidea (Echinodermata).
Bull. Mar. Sci. 46: 701-718.

Blankley,W.O. 1984.
Ecology of the starfish *Anasterias rupicola* at Marion Island (Southern Ocean).
Mar. Ecol. Prog. Ser. 18: 131-138.

Bosch,I. 1989.
Contrasting modes of reproduction in two antarctic asteroids of the genus *Porania*, with a description of unusual feeding and non-feeding larval types. Biol. Bull. 177: 77-82.

Bosch,I. and J.S.Pearse. 1990.
Developmental types of shallow-water asteroids of McMurdo Sound, Antarctica. Mar. Biol. (Berlin) 104: 41-46.

Bouillon,J. and M.Jangoux. 1985.
Note on the relationship between the parasitic mollusc *Thyca crystallina* [Gastropoda, Prosobranchia] and the starfish *Linckia laevigata* [Echinodermata] on Laing Island reef, Papua New-Guineae. Ann. Soc. R. Zool. Belg. 114(2): 249-256.

Bradbury,R.H. and R.Reichelt. 1982.
The reef and man: Rationalising management through ecological theory. Proc. 4th Int. Coral Reef Symp.1: 219-223.

Brahimi-Horn, M.C., Guglielmino, M.L., Sparrow, L.G., Logan, R.I. and P.J. Moran. 1989.
Lipolytic enzymes of the digestive organs of the crown-of-thorns starfish (*Acanthaster planci*): Comparison of the stomach and pyloric caeca.
Comp. biochem. physiol. 92: 637-644.

Brown, J.H. 1981.
Two decades of Homage to Santa Rosalia: Towards a theory of diversity. Amer. Zool. 21: 877-888.

Bruno, I., Minale, L., Riccio, R., Cariello, L., Higa, T. and J. Tanaka. 1993.
Starfish saponins: Part 50. Steroidal glycosides from the Okinawan starfish *Nardoa tuberculata*.
J. natural products 56: 1057-1064.

Bullock, T.H. 1953.
Predator recognition and escape responses of some intertidal gastropods in presence of starfish. Behaviour 5: 130-140.

Burkenroad, M.D. 1957.
Intensity of settling of starfish in Long Island Sound in relationship to fluctuations of the stock of adult starfish and in the settling of oysters. Ecology 38: 164-165.

Cameron, A.M. and R. Endean. 1982.
Renewed population outbreaks of a rare and specialised carnivore (the starfish *Acanthaster planci*) in a complex high-diversity system (the Great Barrier Reef).
Proc. 4th Int. Coral Reef Symp. 2: 593-596.

Cameron, A.M., Endean, R. and L.M. Devantier. 1991.
Predation on massive corals: Are devastating population outbreaks of *Acanthaster planci* novel events?
Mar. ecol. prog. ser. 75: 251-258.

Casapullo, A., Finamore, E., Minale, L., Zollo, F., Carre, J.B.,

Debitus,C., Laurent,D., Folgore,A. and F.Galdiero. 1993.
Starfish saponins: Part 49. New cytotoxic steroidal
glycosides from the starfish *Fromia monilis*.
J. natural products 56: 105-115.

Caswell,H. 1982.
Stable population structure and reproductive value for
populations with complex life cycles. Ecology 63: 1223-1231.

Chao,S.M. and K.H.Chang. 1989.
Some shallow-water asteroids (Echinodermata: Asteroidea) from
Taiwan. Bull. Inst. Zool. Acad. Sinica 28: 215-224.

Charnov,E.L. and W.M.Schaffer. 1973.
Life-History consequences of natural selection: Cole's result
revisited. Am. Nat. 107: 791-793.

Chesher,R. 1969 a.
Destruction of Pacific corals by the sea star *Acanthaster
planci*. Science 165: 280-283.

Chesher,R.H. 1969 b.
Acanthaster planci impact on Pacific coral reefs.
Final Rep. Res. Lab. Westinghouse Elect. Corp. to U.S. Dept.
Interior, 115 pp.

Christensen,A.M. 1970.
Feeding biology of the sea-star *Astropecten irregularis*
(Pennant). Ophelia 8: 1-134.

Clark,A.H. 1931.
Echinoderms from the islands of Niuafoou and Nukualofa,
Tonga Archipelago. Proc. U.S. natn Mus. 80: 1-12.

Clark,A.H. 1952.
Echinoderms from the Marshal Islands.
Proc. U.S. natn Mus. 102: 265-303.

Clark,A.M. 1967 a. Echinoderms from the Red Sea.
Part 2.- Crinoids, Ophiuroids, Echinoids and more Asteroids.
Bull. Sea Fish. Res. Stn Israel. 41: 26-58.

Clark,A.M. 1982.
Echinoderms of Hong Kong.
Proc. Int. Mar. Biol. Workshop 1(1): 485-501.

Clark,A.M. and P.Spencer-Davies. 1966.
Echinoderms of the Maldive Islands.
Ann. Mag. nat. Hist. 13(8): 597-612.

Clark,A.M. and F.W.E.Rowe. 1971.
Monograph of shallow-water Indo-West Pacific Echinoderms.
1-228, pls.1-31, Trustees of the British Museum (Nat. Hist.).

Clark,H.L. 1913.
Autotomy in *Linckia*. Zool. Anz. 42: 156-159.

Clark,H.L. 1921.
The Echinoderm fauna of Torres Strait. Pap. Dep. Mar. Biol.
Carnegie Inst. Wash., 10: viii + 223, pls.1-38.

Clark,H.L. 1938.
Echinoderms from Australia.
Mem. Mus. Comp. Zool. Harv. 55: 1-596.

Clark,H.L. 1946.
The Echinoderm fauna of Australia, its composition and origin.
Carnegie Inst. Wash. 566: 1-567.

Cole,L.C. 1954.
The population consequences of life history phenomena.
Quart. Rev. Biol. 29: 103-137.

Connell,J.H. 1970.
On the role of natural enemies in preventing competitive exclusion in some marine animals and in rainforest trees.
in Dynamics of Populations, den Boer,P.J. and G.R.Gradwells (eds): 298-312.

Connell,J.H. 1978.
Diversity in tropical rain forests and coral reefs.
Science 199: 1302-1310.

Connor,E.F. and E.D.McCoy. 1979.
The statistics and biology of the species-area relationship.
Am. Nat. 113: 791-833.

Connor,E.F. and D.Simberloff. 1979.
The assembly of species communities; chance or competition?
Ecology 60: 1132-1140.

Connor,E.F., E.D.McCoy and B.J.Cosby. 1983.
Model discrimination and expected slope values in species-area studies. Am. Nat. 122: 789-796.

Crump,R.G. and M.F.Barker. 1985.
Sexual and asexual reproduction in geographically separated populations of the fissiparous asteroid *Coscinasterias calamaria*. J. Exp. Mar. Biol. Ecol. 88: 109-128.

Davis,L.V. 1967.
The suppression of autotomy in *Linckia multifora* (Lamarck) by the parasitic gastropod, *Stylifer linckiae* (Sarasin).
Veliger 9: 343-346.

Dayton,P.K., R.J.Rosenthal, L.C.Mahan and T.Antezana 1977.
Population structure and foraging biology of the predacious
Chilean asteroid *Meyenaster gelatinosus* and the escape biology
of its prey. Mar. Biol. 39: 361-370.

De Celis,A.K. 1980.
The asteroids of Marinduque Island, Philippines.
Acta manil. 19: 20-74.

Den Boer,P.J. 1971.
Stabilization of animal numbers and the heterogeneity of the
environment: the problem of the persistence of sparse
populations. in Dynamics of Populations, den Boer,P.J. and
G.R.Gradwells (eds): 77-97.

Denny,M.W. and M.F.Shibata 1989.
Consequences of surf-zone turbulence for settlement and
external fertilisation. Am. Nat. 134: 859-889.

Dight,I.J., Bode,L. and M.K.James. 1990.
Modelling the larval dispersal of *Acanthaster planci*: I.
Large scale hydrodynamics, Cairns Section, Great Barrier Reef
Marine Park (Australia). Coral Reefs 9: 115-123.

Dight,I.J., James,M.K. and L.Bode. 1990.
Modelling the larval dispersal of *Acanthaster planci*: II.
Patterns of reef connectivity. Coral Reefs 9: 125-134.

Doherty,P.J. 1982.
Coral reef fishes: recruitment-limited assemblages.
Proc. 4th Int. Coral reef Symp. 2: 465-470.

Domantay,J.S. 1972.
Monographic studies and checklist of Philippine littoral
Echinoderms. Acta manil. 15: 91-149.

Dubois,P. and M.Jangoux. 1990.
Stereom morphogenesis and differentiation during regeneration of adambulacral spines of *Asterias rubens* (Echinodermata, Asteroidea). Zoomorphology (Berlin) 109: 263-272.

Dunbar,M.J. 1960.
The evolution of stability in marine environments: Natural selection at the level of the ecosystem. Am. Nat. 94: 129-136.

Dundar,M.J. 1972.
The ecosystem as unit of natural selection.
Trans. Conn. Acad. Arts Sci. 44: 113-130.

Dunbar,M.J. 1980.
The blunting of Occam's Razor, or to hell with parsimony.
Can. J. Zool. 58: 123-128.

Ebert,T. 1972.
Estimating growth and mortality rates from size data.
Oecologia 11(3): 281-298.

Ebert,T. 1982.
Longevity, life history and relative body wall size in sea urchins. Ecol. Monogr. 52: 353-394.

Ebert,T. 1983.
Recruitment in echinoderms. pp 169-203 in Echinoderm Studies vol 1. Lawrence.J and M.Jangoux (eds), A.A.Balkema, Rotterdam.

Ebert,T. 1985.
Sensitivity of fitness to macroparameter changes: an analysis of survivorship and individual growth in sea urchin life histories. Oecologia 65: 461-467.

Eckardt,F.E. 1974.
Life-form, survival strategy and CO_2 exchange.
Proc. 1st. Int. Cong. Ecol: 57-59.

Edmondson, C.H. 1935.
Autotomy and regeneration in Hawaiian starfishes.
B.P.Bishop Mus. Occ. Pap. 11(8): 1-29.

Egloff, D.A., Smouse, D.T.Jr. and J.E.Pembroke. 1988.
Penetration of the radial hemal and perihemal systems of *Linckia laevigata* (Asteroidea) by the proboscis of *Thyca crystallina*, an ectoparasitic gastropod. Veliger 30: 342-346.

Elder, H.Y. 1979.
Studies on the host parasite relationship between the parasitic prosobranch *Thyca crystallina* and the asteroid *Linckia laevigata*. J. Zool. 187(3): 369-392.

Ely, C.A. 1942.
Shallow water Asteroidea and Ophiuroidea of Hawaii.
B.P.Bishop Mus. Bull. 176: 1-163.

Emson, R.H. and I.C.Wilkie. 1980.
Fission and autotomy in echinoderms.
Oceanogr. mar. Biol. A. Rev. 18: 155-250.

Endean, R. 1953.
Queensland Faunistic Records Part III. - Echinodermata (excluding Crinoidea). Pap. Dep. Zool. Univ. Qd. 1: 53-60.

Endean, R. 1956.
Queensland Faunistic Records Part IV. - Further records of Echinodermata (excluding Crinoidea).
Pap. Dep. Zool. Univ. Qd. 1: 123-140.

Endean, R. 1957.
The biogeography of Queensland's shallow water echinoderm fauna (excluding Crinoidea) with a rearrangement of the faunistic provinces of tropical Australia.
Aust. J. mar. Freshw. Res., 8(3): 233-273.

Endean, R. 1961.

Queensland faunistic records. Part VII. - Additional records of Echinodermata (excluding Crinoidea).
Pap. Dep. Zool. Univ. Qd. 1: 289-298.

Endean,R. 1965.
Queensland faunistic records. Part VIII. - Further records of Echinodermata (excluding Crinoidea) from southern Queensland.
Pap. Dep. Zool. Univ. Qd. 2: 229-235.

Endean,R. 1969.
Report on investigations made into aspects of the current *Acanthaster planci* (crown of thorns) infestations on certain reefs of the Great Barrier Reef.
Fish. Branch, Qld. Dept. Prim. Ind., Bris. 35 pp.

Endean,R. 1977.
Acanthaster planci infestations of reefs of the Great Barrier Reef. Proc. 3rd Int. Coral Reef Symp. 1: 185-191.

Endean,R. 1982.
Crown-of-thorns starfish on the Great Barrier Reef.
Endeavour 6: 10-14.

Endean,R. and A.M.Cameron. 1990 a.
Trends and new perspectives in coral-reef ecology.
pp 469-492 in Ecosystems of the World vol 25, Coral Reefs, Dubinsky,Z. (ed), Elsevier, New York.

Endean,R. and A.M.Cameron. 1990 b.
Acanthaster planci population outbreaks.
pp 419-437 in Ecosystems of the World vol 25, Coral Reefs, Dubinsky,Z. (ed), Elsevier, New York.

Epel, D. 1991.
How successful is the fertilisation process of the sea urchin egg. pp 51-54 in Biology of the Echinodermata. Yanagisawa, Yasumasu, Suzuki and Motokawa (eds) Balkema, Rotterdam.

Feder, H.M. 1955.
The use of vital stains in marking Pacific coast starfish. Calif. Fish. Game 41:245-246.

Feder, H.M. 1963.
Gastropod defensive responses and their effectiveness in reducing predation by starfish. Ecology 44: 505-512.

Fernandes, L. 1990.
Effect of the distribution and density of benthic target organisms on manta tow estimates of their abundance. Coral Reefs 9: 161-165.

Fernandes, L., Marsh, H., Moran, P.J. and D. Sinclair. 1990.
Bias in manta tow surveys of *Acanthaster planci*. Coral Reefs 9: 155-160.

Fisher, W.K. 1906.
The starfish of the Hawaiian Islands. Bull. U.S. Fish. Commn 23: 987-1130.

Fisher, W.K. 1911.
Asteroidea of the north Pacific and adjacent waters, 1. Bull. U.S. natn Mus. 76: 1-420.

Fisher, W.K. 1919.
Starfishes of the Philippine Sea and adjacent waters. Bull. U.S. natn Mus. 100: 1-711.

Fisher, W.K. 1925.
Sea-stars of the tropical central Pacific. B.P. Bishop Mus. Bull. 27: 63-88.

Fisher,R.A., Corbet,A.S. and C.B.Williams. 1943.
The relation between the number of species and the number of individuals in a random sample of an animal population.
J. anim. ecol. 12: 42-58.

Fisk,D.A. and V.J.Harriott. 1990.
Spatial and temporal variation in coral recruitment on the Great Barrier Reef (Australia): Implications for dispersal hypotheses. Mar. Biol. (Berlin) 107: 485-490.

Frank,P.W. 1968.
Life histories and community stability. Ecology 49: 355-357.

Frank,P.W. 1969.
Growth rates and longevity of some gastropod mollusks on the coral reef at Heron Island. Oecologia (Berlin) 2: 232-250.

Franklin,S.E. 1980.
The reproductive biology and some aspects of the population ecology of the holothurians *Holothuria leucospilota* (Brandt) and *Stichopus chloronotus* (Brandt).
Ph.D. Thesis, University of Sydney.

Gilpin,M.E. and J.M.Diamond. 1982.
Factors contributing to non-randomness in species co-occurrences on islands. Oecologia 52: 75-85.

Gemmill,J.F. 1915.
On the ciliation of asteroids, and on the question of ciliary nutrition in certain species. Proc. zool. Soc. Lond. 1: 1-19.

Gibbs,P.E., C.M.Clark and Clark,A.M. 1976.
Echinoderms from the Great Barrier Reef.
Bull. Br. mus. Nat. Hist. (Zool). 30(4): 103-144.

Glynn, P.W. 1974.
The impact of *A. planci* on corals and coral reefs in the eastern Pacific. Env. Conserv. 1: 295-304.

Glynn, P.W. 1984.
An amphinomid worm predator of the Crown-of-thorns sea star and general predation on asteroids in eastern and western Pacific coral reefs. Bull. Mar. Sci. 35: 54-71.

Glynn, P.W. and D.A. Krupp. 1986.
Feeding biology of a Hawaiian (USA) sea star corallivore, *Culcita novaeguineae*. J. exp. mar. biol. ecol. 96: 75-96.

Goodman, D. 1974.
Natural selection and a cost ceiling on reproductive effort. Am. Nat. 108: 247-268.

Goodman, D. 1975.
The theory of diversity-stability relationships in ecology. Quart. Rev. Biol. 50: 237-266.

Gorshkov, B.A., Gorshkova, I.A., Stonik, V.A. and G.B. Elyakov. 1982. Effect of marine Glycosides on ATPase activity. Toxicon 20(3): 655-658.

Grassle, J.F. 1973.
Variety in coral reef communities. pp 247-270 in Biology and Geology of Coral Reefs 2, Biology 1, ed. O.A. Jones and R. Endean, New York.

Green, G. 1977.
Ecology of toxicity in marine sponges. Mar. Biol. 40: 207-215.

Grosenbaugh, D.A. 1981.
Qualitative assessment of the Asteroids, Echinoids and Holothurians in Yap Lagoon, West Pacific Ocean. Atoll Res. Bull. 0(254): 49-54.

Guille,A. and M.Jangoux. 1978.
Asterides et Ophiurides littorales de la region d'Amboine (Indonesie). Ann. Inst. oceanogr., Paris. 54(1): 47-74.

Hairston,N.G. 1959.
Species abundance and community organisation.
Ecol. 40: 404-416.

Hairston,N.G., D.W.Tinkle and H.M.Wilbur. 1970.
Natural Selection and the parameters of population growth.
J. Wildl. Manag. 34: 681-690.

Hayashi,R. 1938 a.
Sea stars of the Ryukyu Islands.
Bull. biogr. Soc. Japan. 8: 197-222.

Hayashi,R. 1938 b.
Sea stars of the Ogasawara Islands.
Annotnes zool. jap. 17(1): 59-68.

Hayashi,R. 1938 c.
Sea stars of the Caroline Islands.
Palau Trop. Biol. Stat. Stud. 3: 417-446.

Hays,J. Imbrie,J. and N.J.Shackleton. 1976.
Variations in the Earth's Orbit: Pacemaker of the Ice Ages.
Science 194: 1121-1132.

Henderson,J.A and J.S.Lucas. 1971.
Larval development and metamorphoses of *Acanthaster planci* (Asteroidea). Nature 232: 655-657.

Hendler,G. 1975.
Adaptational significance of the patterns of ophiuroid development. Amer. Zool. 15: 691-715.

Hurlbert, S.H. 1971.
The non-concept of species diversity.: a critique and alternative parameters. Ecol. 52: 577-586.

Hutchings, P.A. 1981.
Polychaete recruitment onto dead coral substrates at Lizard Island, Great Barrier Reef, Australia.
Bull. Mar. Sci. 31: 410-423.

Iorizzi, M., Minale, L., Riccio, R., Higa, T. and J. Tanaka. 1991.
Starfish saponins: Part 46. Steroidal glycosides and polyhydroxysteroids from the starfish *Culcita novaeguineae*.
J. natural products 54: 1254-1264.

Iwasaki, K. 1993.
Analyses of limpet defence and predator offence in the field. Mar. Biol. (Berlin) 116: 277-289.

Jacobs, J. 1974.
Diversity, stability and maturity in ecosystems influenced by human activities. Proc. 1st. Int. Cong. Ecol.: 94-95.

Jackson, G.A. and R.R. Strathmann. 1981.
Larval mortality from offshore mixing as a link between precompetent and competent periods of development.
Am. Nat. 118: 16-26.

James, D.B. 1972.
Note on the development of the asteroid *Asterina burtoni* Gray. J. Mar. Biol. Assoc. India 14(2): 883-884.

James, D.B. 1973.
Studies on Indian echinoderms.
J. Mar. Biol. Assoc. India 15(2): 556-559.

James,D.B. and J.S.Pearse. 1969.
Echinoderms from the Gulf of Suez and the northern Red Sea.
J. Mar. Biol. Assoc. India 11(1-2): 75-125.

Jangoux,M. 1972 a.
Les asterides de I'lle d'Inhaca (Mozambique) [Echinodermata, Asteroidea].
Annales Mus. r. Afr. cent. (Ser. 8 Sci. Zool.) 208: 1-50.

Jangoux,M. 1972 b.
Le genre *Neoferdina* Livingstone.
Revue Zool. Bot. afr. 87: 775-794.

Jangoux,M. 1978.
Biological results of the Snellius Expedition XXIX.
Zool. Medd. 52: 287-300.

Jangoux,M. 1980.
Le genre *Leiaster* Peters.
Rev. Zool. afr. 94: 86-110.

Jangoux, 1982.
Food and feeding mechanisms: Asteroidea.
pp 117-159 in Echinoderm Nutrition, Jangoux,M. and J.M.Lawrence (eds), A.A.Balkema, Rotterdam.

Jangoux,M. 1984.
The littoral asteroids from New-Caledonia.
Bull. Mus. Natl. Hist. Nat. 6: 279-294.

Jangoux,M. and A.Aziz. 1985.
The asteroids (Echinodermata) of the central-west part of the Indian ocean (Seychelles, Maldive Archipelagoes).
Bull. Mus. Nat. Hist. Zool. 6: 857-884.

Jell, J. and P. Flood. 1978.
Guide to the geology of reefs of the Capricorn and Bunker Groups, Great Barrier Reef Province, with special reference to Heron Reef. Pap. Dep. Geol. Univ. Qld. 8: 1-85.

Johnson, C.R., Sutton, D.C., Olson, R.R. and R. Giddins. 1991.
Settlement of crown-of-thorns starfish: Role of bacteria on surfaces of coralline algae and a hypothesis for deepwater recruitment. Mar. ecol. prog. ser. 71: 143-162.

Johnson, M.S. and T.J. Threlfall. 1987.
Fissiparity and population genetics of *Coscinasterias calamaria*. Mar. Biol. (Berlin) 93: 517-526.

Jost, P. 1979.
Reaction of two sea star species to an artificial prey patch. Proc. European Colloquium on Echinoderms, Brussels.: 197.

Julka, J.M. and S. Das. 1978.
Studies on the shallow water starfishes of the Andaman and Nicobar Islands. Mitt. Zool. Mus. Berl. 54(2): 345-352.

Kanatani, H. 1969.
Oocyte maturation with 1.Methyl adenine.
Expl. Cell. Res. 57: 333-337.

Kanatani, H. 1973.
Maturation-inducing substance in starfishes.
Int. Rev. Cytol. 35: 253-298.

Keesing, J.K. and A.R. Halford. 1992.
Field measurement of survival rates of juvenile *Acanthaster planci*: Techniques and preliminary results. Mar. ecol. prog. ser. 85: 107-114.

Keesing,J.K. and J.S.Lucas 1992.
Field measurement of feeding and movement rates of the crown-of-thorns starfish *Acanthaster planci* (L.).
J. Exp. Mar. Biol. Ecol. 156: 89-104.

Keesing,J.K. and C.M.Cartwright. 1993.
Measuring settlement intensity of echinoderms on coral reefs. Mar. biol. (Berlin) 117: 399-407.

Kenchington,R.A. 1976.
Acanthaster planci on the Great Barrier Reef: detailed surveys of four transects between 19° and 20°S.
Biol. Conserv. 9: 165-179.

Kerr,A.M., Norris,D.R., Schupp,P.J., Meyer,K.D., Pitlik,T.J., Hopper,D.R., Chamberlain,J.D. and L.S.Meyer. 1992.
Range extensions of echinoderms (Asteroidea, Echinoidea and Holothuroidea) to Guam, Mariana Islands.
Micronesica 25: 201-216.

Kicha,A.A., Kalinovskii,A.I. and E.V.Levina. 1985.
Culcitoside C-1 from the starfishes *Culcita novaeguineae* and *Linckia guildingi*.
Khimiya Prirodnykh Soedinenii 0 (6): 801-804.

Klopfer,P.H. 1959.
Environmental determinants of faunal diversity.
Am. Nat. 93: 337-342.

Klopfer,P.H. and R.H.MacArthur. 1960.
Niche size and faunal diversity. Am. Nat. 94: 293-300.

Klumpp,D.W. and A.Pulfrich. 1989.
Trophic significance of herbivorous macroinvertebrates on the central Great Barrier Reef (Australia). Coral Reefs 8:135-144.

Koehler,R. 1910.
Shallow water Asteroidea.
Echinoderms of the Indian Museum 6. Calcutta, 192 pp.

Kohn,A.J. 1959.
The ecology of *Conus* in Hawaii. Ecol. Monogr. 29: 47-90.

Kohn,A.J. 1968.
Microhabitats, abundance and food of *Conus* on atoll reefs in the Maldive and Chagos Islands. Ecology 49: 1046-1062.

Kohn,A.J. and P.J.Leviten. 1976.
Effect of habitat complexity on population density and species richness in tropical intertidal predatory gastropod assemblages. Oecologia 25: 199-210.

Komatsu,M. 1973.
A preliminary report on the development of the sea-star *Leiaster leachi*. Proc. Jap. Soc. Syst. Zool. 9: 55-58.

Komatsu,M., Kano,Y.T. and C.Oguro. 1990.
Development of a true ovoviviparous sea star, *Asterina pseudoexigua pacifica* Hayashi. Biol. Bull. 179: 254-263.

Kuborta,J., Nakao,K., Shirai,H. and H.Kanatani. 1977.
1.Methyl adenine-producing cells in the starfish testes. Exp. Cell. Res. 106: 63-70.

Kunin,W.E. and K.J.Gaston. 1993.
The biology of rarity: Patterns, causes and consequences. Tree 8(8): 298-301.

Kwon,W.S. and C.H.Cho. 1986.
Culture of the ark shell, *Anadara broughtonii* in Yoja Bay (Korea). Bull. Korean Fish. Soc. 19: 375-379.

Lawrence,J.M., Klinger,T.S., McClintock,J.B., Watts,S.A., Chen,C.P., Marsh,A. and L.Smith. 1986.

Allocation of nutrient resources to body components by regenerating (*Luidia clathrata*) (Echinodermata: Asteroidea). J. exp. mar. biol. ecol. 102: 47-54.

Laxton, J.H. 1971.
Feeding in some Australian Cymatiidae (Gastropoda: Prosobranchia). Zool. J. Linn. Soc. 50: 1-9.

Laxton, J.H. 1974.
A preliminary study of the biology and ecology of the blue starfish *Linckia laevigata* (L) on the Australian Great Barrier Reef and an interpretation of its role in the coral reef ecosystem. Biol. J. Linn. Soc. 6: 47-64.

Leigh, E.G. 1965.
On the relationship between productivity, biomass, diversity, and stability of a community.
Proc. Nat. Acad. Sci. 53: 777-783.

Lessios, H.A. 1990.
Adaptation and phylogeny as determinants of egg size in echinoderms from the two sides of the Isthmus of Panama.
Am. Nat. 135: 1-13.

Levins, R. and Culver. 1971.
Regional coexistence of species and competition between rare species. Proc. Nat. Acad. Sci. 68: 1246-1248.

Leviten, P.J. and A.J. Kohn. 1980.
Microhabitat resource use, activity patterns, and episodic catastrophe: *Conus* on intertidal reef rock benches.
Ecol. Monogr. 50: 55-75.

Liao,Y. 1980.
The echinoderms of Xisha Islands, Guangdong Province, China. 4. Asteroidea. Studia Mar. Sin. 17: 153-171.

Livingstone,A.A. 1932.
Asteroidea. Sci. Rept, G.B.R. Exped. 4(8): 241-265.

Loosanoff,V.L. 1937.
Use of Nile Blue Sulphate in marking starfish. Science 85(2208): 412.

Loosanoff,V.L. 1961.
Biology and methods of controlling the starfish, *Asterias forbesi*. Fishery Leaflet 520, US Department of Interior, Washington DC.

Loosanoff,V.L. 1964.
Variation in time and intensity of settling of the starfish, *Asterias forbesi*, in Long Island Sound during a twenty-five year period. Biol. Bull. 126: 423-439.

Lucas,J. 1984.
Growth, maturation and effects of diet in *Acanthaster planci* (Asteroidea) and hybrids reared in the laboratory. J. exp. Mar. Biol. Ecol. 79: 129-148.

MacArthur,R.H. 1955.
Fluctuations of animal populations, and a measure of community stability. Ecol. 36: 533-536.

MacArthur,R.H. and R.Levins. 1964.
Competition, habitat selection and character displacement in a patchy environment. Proc. Nat. Acad. Sci. 51: 1207-1210.

Macarthur,R.H. and R.Levins. 1967.
The limiting similarity, convergence and divergence of coexisting species. Am. Nat. 101: 377-385.

Margalef,R. 1963.
On certain unifying principles in ecology.
Am. Nat. 97: 357-374.

Margalef,R. 1974.
Diversity, stability and maturity in natural ecosystems.
Proc. 1st. Int. Cong. Ecol.: 66.

Marsh,L.M. 1974.
Shallow-water Asterozoans of southeastern Polynesia
1. Asteroidea. Micronesica 10: 65-104.

Marsh,L.M. 1976.
West Australian Asteroidea since H.L.Clark.
Thalassia Jugoslavica 12(1): 213-225.

Marsh,L.M. 1977.
Coral Reef Asteroids of Palau, Caroline Islands.
Micronesica 13(2): 251-281.

Marsh,L.M. 1991.
A revision of the echinoderm genus *Bunaster* (Asteroidea: Ophidiasteridae). Rec. West. Aust. Mus. 51: 419-434.

Martin,T.E. 1981.
Species-area slopes and coefficients: a caution on their interpretation. Am. Nat. 118: 823-837.

Mauzey,K.P., C.Birkeland and P.K.Dayton 1968.
Feeding behaviour of asteroids and escape responses of their prey in the Puget Sound region. Ecology 49: 603-619.

Maxwell,W.G.H. 1968.
Atlas of the Great Barrier Reef. Elsevier, New York.

May,R.M. 1972.
Will a large complex system be stable? Nature 238: 413-414.

May,R.M. and R.H.MacArthur. 1972.
Niche overlap as a function of environmental variability.
Proc. Nat. Acad. Sci. 69: 1109-1113.

McCallum,H.I. 1987.
Predator regulation of *Acanthaster planci*.
J. theor. biol. 127: 207-220.

McCallum,H.I., Endean,R. and A.M.Cameron. 1989.
Sublethal damage to *Acanthaster planci* as an index of predation pressure. Mar. ecol. prog. ser. 56: 29-36.

McClary,D.J. and P.V.Mladenov. 1989.
Reproductive pattern in the brooding and broadcasting sea star *Pteraster militaris*. Mar. Biol. (Berlin) 103: 531-540.

McClary,D.J. 1990.
Brooding biology of the sea star *Pteraster militaris* (O.F. Mueller): Energetic and histological evidence for nutrient translocation to brooded juveniles.
J. mar. biol. ecol. 142: 183-200.

McEdward,L.R. and F.S.Chia. 1991.
Size and energy content of eggs from echinoderms with pelagic lecithotrophic development.
J. exp. mar. biol. ecol. 147: 95-102.

McEdward,L.R. and D.A.Janies. 1993.
Life cycle evolution in asteroids: What is a larva?
Biol. Bull. 184: 255-268.

McGuinness,K.A. 1984.
Equations and explanations in the study of species-area curves. Biol. Rev. 59: 423-440.

Mead, A.D. 1900.
On the correlation between growth and food supply in the starfish. Am. Nat. 34: 17-23.

Menge, B.A. 1972 a.
Foraging strategy of a starfish in relation to actual prey availability and environmental predictability.
Ecol. Monogr. 42: 25-50.

Menge, B.A. 1972 b.
Competition for food between two intertidal starfish species and its effect on body size and feeding. Ecology 53: 635-644.

Menge, B.A. 1975.
Brood or broadcast? The adaptive significance of different reproductive strategies in two intertidal sea-stars *Leptasterias hexactis* and *Pisaster ochraceus*.
Mar. Biol. 31: 87-100.

Menge, B.A. 1981.
Effects of feeding on the environment: Asteroidea.
pp 521-551 in Echinoderms Nutrition, Jangoux, M. and J.M. Lawrence eds., A.A. Balkema, Rotterdam.

Mileikovsky, S.A. 1971.
Types of larval development in marine bottom invertebrates, their distribution and ecological significance:
A reevaluation. Mar. Biol. 10: 193-213.

Miller, R.L. 1989.
Evidence for the presence of sexual pheromones in free-spawning starfish. J. exp. mar. biol. ecol. 130: 205-222.

Minale, L., Pizza, C., Riccio, R., Zollo, F., Pusset, J. and P. Laboute 1984. Starfish saponins 13. Occurrence of nodososide in the starfish *Acanthaster planci* and *Linckia laevigata*.
J. Nat. Prod. 47(3): 558.

Minchin,D. 1987.
Sea-water temperature and spawning behavior in the sea star *Marthasterias glacialis*. Mar. Biol. (Berlin) 95: 139-144.

Miyazawa,K., Noguchi,T., Maruyama,J., Jeon,J.K., Otsuka,M. and K.Hashimoto. 1985.
Occurrence of tetrodotoxin in the starfishes *Astropecten polyacanthus* and *Astropecten scoparius* in the Seto Inland Sea. Mar. Biol. (Berlin) 90: 61-64.

Miyazawa,K., Higashiyama,M., Hori,K., Noguchi,T., Ito,K. and K.Hashimoto. 1987.
Distribution of tetrodotoxin in various organs of the starfish *Astropecten polyacanthus*. Mar. Biol. (Berlin) 96: 385-390.

Mladenov,P.V. and R.H.Emson. 1984.
Divide and broadcast: Sexual reproduction in the West Indian brittle star *Ophiocomella ophiactoides* and its relationship to fissiparity. Mar. Biol. (Berlin) 81: 273-282.

Mladenov,P.V., Carson,S.F. and C.W.Walker. 1986.
Reproductive ecology of an obligately fissiparous population of the sea star *Stephanasterias albula*.
J. exp. mar. biol. ecol. 96: 155-176.

Mladenov,P.V., Bisgrove,B., Asotra,S. and R.D.Burke. 1989.
Mechanisms of arm-tip regeneration in the sea star, *Leptasterias hexactis*. Roux's arch. devel. biol. 198: 19-28.

Mladenov,P.V. and R.H.Emson. 1990.
Genetic structure of populations of two closely related brittle stars with contrasting sexual and asexual life histories, with observations on the genetic structure of a second asexual species. Mar. Biol. (Berl) 104: 265-274.

Moran, P. 1986.
The *Acanthaster* phenomenon.
Oceanogr. Mar. Biol. Ann. Rev. 24: 379-480.

Moran, P.J. and G. De'ath. 1992 a.
Estimates of the abundance of the crown-of-thorns starfish *Acanthaster planci* in outbreaking and non-outbreaking population on reefs within the Great Barrier Reef.
Mar. biol. (Berlin) 113: 509-515.

Moran, P.J. and G. De'ath. 1992 b.
Suitability of the manta tow technique for estimating relative and absolute abundances of crown-of-thorns starfish (*Acanthaster planci* L.) and corals.
Aust. J. mar. freshw. res. 43(2): 357-378.

Morse, D.E. 1984.
Biochemical control of larval recruitment and marine fouling.
pp 134-141 in Marine Biodeterioration: An Interdisciplinary Study. Costlow, J.D. and R.C. Tipper eds., Naval Inst. Press, Annapolis, Maryland.

Mortensen, T. 1937.
Contributions to the study of the development and larval forms of echinoderms. III.
K. danske Vidensk. Selsk. Skr. 9 Raekke 7(1): 1-65.

Mortensen, T. 1938.
Contributions to the study of the development and larval forms of echinoderms. IV.
K. danske Vidensk. Selsk. Skr. 9 Raekke 7(3): 1-59.

Mortensen, T. 1940.
Echinoderms from the Iranian Gulf. Asteroidea.
Dan. scient. Invest. Iran. 2: 55-110.

Muenchow,G. 1978.
A note on the timing of sex in asexual/sexual organisms.
Am. Nat. 112: 774-779.

Murphy,G.I. 1968.
Pattern in life-history and the environment.
Am. Nat. 102: 391-411.

Narita,H., Nara,M., Baba,K., Ohgami,H., T.K.Ai., Noguchi,T. and K.Hashimoto. 1984.
Effect of feeding a trumpet shell, *Charonia sauliae* with toxic starfish (*Astropecten polyacanthus*).
J. Food Hygienic Soc. Japan 25: 251-255.

Nash,W.J., Goddard,M. and J.S.Lucas. 1988.
Population genetic studies of the crown-of-thorns starfish, *Acanthaster planci* (L.), in the Great Barrier Reef region (Australia). Coral Reefs 7: 11-18.

Newell,N.D. 1972.
The evolution of reefs. Sci. Am. 226 (6): 54-65.

Nishida,M. and J.S.Lucas. 1988.
Genetic differences between geographic populations of the crown-of-thorns starfish throughout the Pacific region.
Mar. Biol. (Berlin) 98: 359-368.

Noguchi *et al*. 1982.
Tetrodotoxin in the starfish *Astropecten polyacanthus* in association with toxification of a trumpet shell, "Boshubora" *Charonia sauliae*. Bull. Jap. Soc. Sci. Fish. 48: 1173-1177.

Noguchi,T., Sakai,T., Maruyama,J., Jeon,J.K., Kesamaru,K. and K.Hashimotu. 1985 a.
Toxicity of a trumpet shell, *Charonia sauliae* ("Boshubora") inhabiting along the coasts of Miyazaki Prefecture (Japan).
Bull. Jap. Soc. Scient. Fish. 51: 677-680.

Noguchi,T., Jeon,J.K., Maruyama,J., Sato,Y., Saisho,T. and K.Hashimoto. 1985 b.
Toxicity of trumpet shells inhabiting the coastal waters of Kagoshima prefecture (Japan) along with identification of the responsible toxin.
Bull. Jap. Soc. Scient. Fish. 51: 1727-1732.

Nojima,S., Soliman,F.E.S., Kondo,Y., Kuwano,Y., Nasu,K. and C. Kitajima. 1986.
Some notes on the outbreak of the sea star, *Asterias amurensis versicolor*, in the Ariake Sea, western Kyushu (Japan).
Pub. Amakusa Mar. Biol. Lab. 8: 89-112.

Oguro,C. 1984.
Supplementary notes on the sea-stars from the Palau and Yap Islands 1. Annot. Zool. Jpn 56(3): 221-226.

Oguro,C., Komatsu,M. and Y.T.Kano. 1975.
A note on the early development of *Astropecten polyacanthus* (M&T). Proc. Jap. Soc. Syst. Zool. 11: 49-52.

Okaji,K. 1991.
Delayed spawning activity in dispersed individuals of *Acanthaster planci* in Okinawa. pp 291-295 in Biology of the Echinodermata. Yanagisawa, Yasumasu, Suzuki and Motokawa (eds) Balkema, Rotterdam.

Olsen,R.R. 1987.
In situ culturing as a test of the larval starvation hypothesis for the crown-of-thorns starfish, *Acanthaster planci*. Limnol. oceanogr. 32: 895-904.

Ormond, R.F.G. and A.C.Campbell. 1973.
Formation and breakdown of *Acanthaster planci* aggregations in the Red Sea. Proc. 2nd Int. Coral Reef Symp. 1: 595-619.

Ormond,R.F.G., N.J.Hanscomb and D.H.Beach. 1976.
Food selection and learning in the crown-of-thorns starfish, *Acanthaster planci*. Mar. Behav. Physiol. 4(2): 93-105.

Ottesen,P.O. and J.S.Lucas. 1982.
Divide or Broadcast: Interrelation of Asexual and Sexual Reproduction in a Population of the Fissiparous Hermaphroditic Seastar *Nepanthia belcheri* (Asteroidea: Asterinidae). Mar. Biol. 69: 223-233.

Patton,M.L., Brown,S.T., Harman,R.F. and R.S.Grove. 1991.
Effect of the anemone *Corynactis californica* on subtidal predation by sea stars in the southern California Bight (USA). Bull. mar. sci. 48: 623-634.

Pearse,J.S. 1968.
Patterns of reproductive periodicities in four species of Indo-Pacific echinoderms. Proc. Indian Acad. Sci. 67: 247-279.

Pearse,J.S. 1970.
Reproductive periodicities if Indo-Pacific invertebrates in the Gulf of Suez. 3. The echinoid *Diadema setosum* (Leske). Bull. Mar. Sci. 20: 697-720.

Pearse,J.S. 1975.
Lunar reproductive rhythms in sea urchins. A review. J. Interdiscipl. Cycle Res. 6: 47-52.

Pearson,R.G. and R.Endean. 1969.
A preliminary study of the coral predator *Acanthaster planci* (L.) (Asteroidea) on the Great Barrier Reef. Fisheries Notes Qld. Dept. Harbours and Marine, Brisbane 3: 27-55.

Pennington, J.T. 1985.
The ecology of fertilisation of echinoid eggs: the consequences of sperm dilution, adult aggregation, and synchronous spawning. Biol. Bull. 169: 417-430.

Percharde, P.L. 1972.
Observations on the gastropod *Charonia variegata*, in Trinidad and Tobago. Nautilus, Philad. 85: 84-92.

Peters, R.H. 1976.
Tautology in evolution and ecology. Am. Nat. 110: 1-12.

Phillips, D.W. 1976.
The effect of a species-specific avoidance response to predatory starfish on the intertidal distribution of two gastropods. Oecologia (Berlin) 23: 83-94.

Pianka, E.R. 1966.
Latitudinal gradients in species diversity: a review of concepts. Am. Nat. 100: 33-46.

Pianka, E.R. 1972.
r- and K- selection or b- and d- selection?
Am. Nat. 106: 581-588.

Pielou, E.C. 1981.
The usefulness of ecological models: A stocktaking.
Quart. Rev. Biol. 56: 17-31.

Pimm, S.L. 1984.
The complexity and stability of ecosystems.
Nature 307: 321-326.

Pope, E.C. and F.W.E. Rowe. 1977.
A new genus and two new species in the family Mithrodiidae [Echinodermata, Asteroidea] with comments on the status of the *Mithrodia* species. Aust. Zool. 19: 201-216.

Price, A.R.G. 1981.
Echinoderm fauna of the western Arabian Gulf.
J. Nat. Hist. 15: 1-16.

Price, A.R.G. 1982.
Western Arabian Gulf echinoderms in high salinity waters and the occurrence of dwarfism. J. Nat Hist. 16(4): 519-528.

Quinn, J.F. and A.E. Dunham. 1983.
On hypothesis testing in ecology and evolution.
Am. Nat. 122: 602-617.

Reichelt, R.E. 1982.
Space: A non-limiting resource in the niches of some abundant coral reef gastropods. Coral Reefs 1: 3-11.

Ribi, G. and P. Jost, 1978.
Feeding rate and duration of daily activity of *Astropecten aranciacus* (Echinodermata: Asteroidea) in relation to prey density. Marine Biology 45: 249-254.

Riccio, R., Dini, A., Minale, L., Pizza, C., Zollo, F. and T. Sevenet 1982. Starfish saponins VII. Structure of luzonicoside, a further steroidal cyclic glycoside from the Pacific starfish *Echinaster luzonicus*.
Experimentia (Basel) 38: 68-70.

Riccio, R., Greco, O.S., Minale, L., Pusset, J. and J.L. Menou 1985. Starfish saponins 18. Steroidal glycoside sulphates from the starfish *Linckia laevigata*. J. Nat. Prod. 48(1): 97-101.

Rideout, R.S. 1975.
Toxicity of the asteroid *Linckia laevigata* (L.) to the damselfish *Dascyllus aruanus* (L.). Micronesica 11(1): 153-154.

Rideout, R.S. 1978. Asexual reproduction as a means of population maintenance in the coral reef asteroid *Linckia multifora* on Guam. Mar. Biol. 47(3): 287-296.

Roff, D.A. 1981.
Reproductive uncertainty and the evolution of iteroparity: Why don't flatfish put all their eggs in one basket?
Can. J. Fish. aquat. Sci. 38: 968-977.

Rothschild, Lord and M.M. Swann. 1951.
The fertilisation reaction in the sea urchin. The probability of a successful sperm-egg collision.
J. exp. biol. 28: 403-416.

Roughgarden, J. 1983.
Competition and theory in community ecology.
Am. Nat. 122: 583-601.

Rowe, F.W.E. 1977.
The status of *Nardoa* subgenus *Andora* [Asteroidea, Ophidiasteridae] with the description of 2 new subgenera and 3 new species. Rec. Aust. Mus. 31(6): 235-244.

Run, J.Q., Chen, C.P., Chang, K.H. and F.S. Chia. 1988.
Mating behavior and reproductive cycle of *Archaster typicus* (Echinodermata: Asteroidea). Mar. Biol. (Berl) 99: 247-254.

Sale, P.F. 1974.
Mechanisms of co-existence in a guild of territorial fishes at Heron Island. Proc. 2nd Int. Coral reef Symp. 1: 193-206.

Sale, P.F. 1976.
Reef fish lottery. Nat. Hist. 85: 60-65.

Sale, P.F. 1977.
Maintenance of high diversity in coral reef fish communities.
Am. Nat. 111: 337-359.

Sale, P.F. 1984.
The structure of communities of fish on coral reefs and the merit of a hypothesis testing, manipulative approach to ecology. in Ecological Communities: Conceptual issues and the evidence. pp 478-490. ed. D.Strong, D.Simberloff, L.Abele and A.Thistle. Princeton Univ. Press, Princeton, N.J.

Sale, P.F. 1991.
Reef fish communities: open nonequilibrial systems. pp 564-598 in The ecology of fishes on coral reefs. P.F.Sale. (ed) Academic Press, San Diego.

Sale, P.F. and R.Dybdahl. 1975.
Determinants of community structure for coral reef fishes in an experimental habitat. Ecology 56: 1343-1355.

Sale, P.F. and W.A.Douglas. 1984.
Temporal variability in the community structure of fish on coral patch reefs and the relation of community structure to reef structure. Ecology 65: 409-422.

Schaffer, W.M. 1974.
Optimal reproductive effort in fluctuating environments. Am. Nat. 108: 783-790.

Scheibling, R.E. 1980.
Dynamics and feeding activity of high-density aggregations of *Oreaster reticulatus* (Echinodermata: Asteroidea) in a sand patch habitat. Mar. Ecol. Prog. Ser. 2: 321-327.

Scheibling, R.E. 1981 a.
Growth and respiration rate of juvenile *Oreaster reticulatus* (L.) (Echinodermata: Asteroidea) on fish and algal diets. Comp. Biochem. Physiol. 69A: 175-176.

Scheibling,R.E. 1981 b.
Optimal foraging movements of *Oreaster reticulatis* (L) (Echinodermata: Asteroidea).
J. Exp. Mar. Biol. Ecol. 51: 173-185.

Scheibling,R.E. 1982.
Feeding habits of *Oreaster reticulatus* (Echinodermata: Asteroidea). Bull. Mar. Sci. 32: 504-510.

Scheltema,R.S. 1968.
Dispersal of larvae by equatorial ocean currents and its importance to the zoogeography of shoal-water tropical species. Nature 217: 1159-1162.

Scheltema,R.S. 1971.
Larval dispersal as a means of genetic exchange between geographically separated populations of shallow water benthic marine gastropods. Biol. Bull. 140: 284-322.

Schmitt,R.J. 1982.
Consequences of dissimilar defences against predation in a subtidal marine community. Ecology 63: 1588-1601.

Shiomi,K., Yamamoto,S., Yamanaka,H. and T.Kikuchi. 1988.
Purification and characterization of a lethal factor in venom from the crown-of-thorns starfish (*Acanthaster planci*). Toxicon 26: 1077-1084.

Shiomi,K., Yamamoto,S., Yamanaka,H., Kikuchi,T. and K.Konno. 1990.
Liver damage by the crown-of-thorns starfish (*Acanthaster planci*) lethal factor. Toxicon 28: 469-476.

Slattery,M. and I.Bosch. 1993.
Mating behavior of a brooding Antarctic asteroid, *Neosmilaster georgianus*. Invert. repro. devel. 24: 97-102.

Sloan,N.A. 1980.

Aspects of the feeding biology of asteroids.
Oceanogr. mar. Biol. ann. Rev. 18: 57-124.

Stearns, S.C. 1977.
Life history tactics: a review of the ideas.
Quart. Rev. Biol. 51: 3-47.

Stevenson, J.P. 1992.
A possible modification of the distribution of the intertidal seastar *Patiriella exigua* (Lamarck) (Echinodermata: Asteroidea) by *Patiriella calcar* (Lamarck).
J. exp. mar. biol. ecol. 155: 41-54.

Strathmann, R.R. 1974.
The spread of sibling larvae of sedentary marine invertebrates. Am. Nat. 108: 29-44.

Strathmann, R.R. 1978.
The evolution and loss of feeding larval stages of marine invertebrates. Evolution 32: 894-906.

Strathmann, R.R. and K. Vedder. 1977.
Size and organic content of eggs of echinoderms and other invertebrates as related to developmental strategies and egg eating. Mar. Biol. 39(4): 305-309.

Strong, R.D. 1975.
Distribution, morphometry, and thermal stress studies on two forms of *Linckia* (Asteroidea) on Guam.
Micronesica 11: 167-183.

Stump, R.J.W. and J.S. Lucas. 1990.
Linear growth in spines from *Acanthaster planci* (L.) involving growth lines and periodic pigment bands.
Coral Reefs 9: 149-154.

Sughihara, G. 1981.
$S=CAZ$, $Z=1/4$: a reply to Connor and McCoy.

Am. Nat. 117: 790-793.

Talbot,F.H., Russell,B.C. and G.R.V.Anderson. 1978.
Coral reef fish communities: unstable high-diversity systems?
Ecol. monogr. 48: 425-440.

Thandar,A.S. 1989.
Zoogeography of the southern African echinoderm fauna.
South Afr. J. zool. 24: 311-318.

Thomassin,B.A. 1976.
The feeding behaviour of the felt-, sponge-, and coral-feeding sea stars, mainly *Culcita schmideliana*.
Helg. wiss. Meeres. 28: 51-65.

Thompson,G.B. and C.Thompson. 1982.
Movement and size structure in a population of the blue starfish *Linckia laevigata* (L.) at Lizard Island, Great Barrier Reef. Aust. J. Mar. Freshw. Res. 33: 561-573.

Thorson,G. 1950.
Reproductive and larval ecology of marine invertebrates.
Biol. Rev. 25: 1-45.

Thorson,G. 1966.
Some factors influencing the recruitment and establishment of marine benthic communities. Neth. J. Sea. Res. 3: 267-293.

Tokeshi,M. 1991.
Extraoral and intraoral feeding: Flexible foraging tactics in the South American sun-star, *Heliaster helianthus*.
J. zool. (London) 225: 439-448.

Tortonese, E. 1960.
Echinoderms from the Red Sea. 1. Asteroidea.
Bull. Sea. Fish. Res. Stn. Israel 29: 17-23.

Tortonese, E. 1977.
Report on echinoderms from the Gulf of Aqaba (Red Sea).
Monit. Zool. Ital. Suppl. 9 (12): 273-290.

Tortonese, E. 1979.
Echinoderms collected along the eastern shore of the Red Sea.
Atti. Soc. Ital. Sci. Nat. Mus. Civ. Stor. Nat. Milano 120: 314-319.

Tortonese, E. 1980.
Researches on the coast of Somalia: Littoral Echinodermata.
Monit. Zool. Ital. Suppl. 13(5): 99-140.

Turner, R.L. 1976.
Sexual difference in latent period of spawning following injection of the hormone 1. Methyl adenine in *Echinaster* (Echinodermata: Asteroidea).
General comp. Endocr. 28: 109-112.

Vance, R.R. 1973.
On reproductive strategies in marine benthic invertebrates.
Am. Nat. 107: 339-352 and 353-361.

Vermeij, G.J. 1987.
Evolution and escalation. Princeton Univ. Press, Princeton, New Jersey. 528pp.

Vernon, A.A. 1937.
Starfish stains. Science 86: 64.

Walenkamp, J.H.C. 1990.
Systematics and zoogeography of Asteroidea (Echinodermata) from Inhaca Island, Mozambique. Zool. Verh. 0(261): 3-86.

Weiss,P. 1969.
This living system: determination stratified.
pp 3-53 in Beyond Reductionism. Koestler,A. and A.Smythies (eds), Hutchinson, London.

Williams,S.T. and J.A.H.Benzie. 1993.
Genetic consequences of long larval life in the starfish *Linckia laevigata* (Echinodermata: Asteroidea) on the Great Barrier Reef. Mar. Biol. (Berlin) 117: 71-77.

Wolanski,E. 1993.
Facts and numerical artefacts in modelling the dispersal of crown-of-thorns starfish larvae in the Great Barrier Reef. Aust. J. mar. freshw. res. 44: 427-436.

Wolda,H. 1970.
Ecological variation and its implications for the dynamics of populations of the land snail *Cepacea nemoralis*. pp 98-108 in Dynamics of Populations. den Boer,P.J. and G.R.Gradwells (eds), Centre for Agricultural Publishing and Documentation, Wageningen.

Yamaguchi,M. 1973 a.
Recruitment of coral reef asteroids, with emphasis on *Acanthaster planci*. Micronesica 9: 207-212.

Yamaguchi,M. 1973 b.
Early life histories of coral reef asteroids, with special reference to *Acanthaster planci*.
pp 369-387 in Biology and Geology of Coral Reefs, 2. Biol 1. O.A.Jones and R.Endean (eds), Academic Press, New York.

Yamaguchi,M. 1974.
Larval life span of the coral reef asteroid *Gomophia egyptiaca* (Gray). Micronesica 10: 57-64.

Yamaguchi,M. 1975 a.
Estimating growth parameters from growth data. Oecologia (Berlin) 20: 321-332.

Yamaguchi,M. 1975 b.
Coral reef asteroids of Guam. Biotropica 7: 12-23.

Yamaguchi,M. 1977 a.
Population structure, spawning, and growth of the coral reef asteroid *Linckia laevigata* (Linnaeus). Pac. Sci. 31: 1330.

Yamaguchi,M. 1977 b.
Larval behaviour and geographic distribution of coral reef asteroids in the Indo-West Pacific. Micronesica 13: 283-296.

Yamaguchi,M. 1977 c.
Estimating the length of the exponential growth phase growth increment observations on the coral reef asteroid *Culcita novaeguineae*. Mar. Biol. 39: 57-60.

Yamaguchi,M. and J.S.Lucas. 1984.
Natural parthenogenesis, larval and juvenile development, and geographical distribution of the coral reef asteroid *Ophidiaster granifer*. Mar. Biol. 83: 33-42.

Zagalsky,P.F., Haxo,F., Hertzberg,S. and S.Liaaen-Jensen. 1989.
Studies on a blue carotenoprotein, linckiacyanin, isolated from the starfish *Linckia laevigata* (Echinodermata: Asteroidea). Comp. biochem. physiol. 93: 339-354.

Zann,L., Brodie,J., Berryman,C. and M.Naqasima. 1987.
Recruitment, ecology, growth and behavior of juvenile *Acanthaster planci* (L.) (Echinodermata: Asteroidea). Bull. Mar. Sci. 41: 561-575.

www.ingramcontent.com/pod-product-compliance
Lightning Source LLC
Chambersburg PA
CBHW060532010526
44107CB00059B/2619